中国国家博物馆
The National Museum of
China in Beijing

主编

(德) 曼哈德·冯·格康
Meinhard von Gerkan
(德) 斯特凡·胥茨
Stephan Schütz
马立东
Ma Lidong

辽宁科学技术出版社
·沈阳·

图书在版编目（CIP）数据

中国国家博物馆 / （德）格康，（德）胥茨，马立东
主编. 一 沈阳：辽宁科学技术出版社，2014.3
    ISBN 978-7-5381-8526-3

    Ⅰ.①中… Ⅱ.①格… ②胥… ③马… Ⅲ.①博物馆-
建筑设计-中国 Ⅳ.①TU242.5

    中国版本图书馆CIP数据核字(2014)第042193号

主　　编：曼哈德·冯·格康（gmp）、斯特凡·胥茨(gmp)、马立东（CABR）
总 策 划：迈克尔·库恩（gmp）、柳青（UED）
策　　划：克劳迪娅·苔斯勒（gmp）、方小诗（gmp）、郑珊珊（gmp）
版式设计：汤姆·魏伯伦茨、亨德里克·西什莱、欧恩平面设计公司（德国汉堡）
封面设计：欧恩平面设计公司（德国汉堡）
公司地址：德国汉堡易北大道139,22763
网　　址：www.gmp-architekten.de www.uedmagazine.net

出版发行：辽宁科学技术出版社
　　　　　（地址：沈阳市和平区十一纬路29号 邮编：110003）
印 刷 者：北京雅昌彩色印刷有限公司
幅面尺寸：190mm×292mm
印　张：6.5
插　页：4
字　数：20千字
出版时间：2014年6月第1版
印刷时间：2014年6月第1次印刷
责任编辑：姜思琪
美术编辑：吴　娜
责任校对：王玉宝

书　号：ISBN 978-7-5381-8526-3
定　价：149.00元

中国国家博物馆

The National Museum
of China in Beijing

gmp·德国冯·格康，玛格及合伙人建筑师事务所
CABR·中国建筑科学研究院

目录

北京

北京作为城市的历史可以追溯到3 000年前。秦汉以来，北京地区一直是华夏民族北方的军事和商业重镇，明成祖朱棣正式迁都燕京，以之为京师，称为"北京"。金陵应天府则作为留都，称南京。1949年，中华人民共和国中央人民政府在北京宣告成立。

北京市面积16 808km²。至2012年约有人口两千万。城市中心为紫禁城宫殿群和天安门广场，广场为人民大会堂、毛主席纪念堂以及中国国家博物馆围合。五条环线交通干道联系了市内各个城区。

北京作为文化历史名城，同时也是中国重要的工业和科研学术中心。

北京相关数据：
面积：16 808km²
人口：2 0693 万

Beijing

In English, the name Beijing means "Northern Capital". During the Ming Dynasty, the Chinese emperors ruled the country from two centers, namely Beijing and Nanjing, the "Southern Capital". Beijing has a 3,000-year history and has always been an important hub of Chinese politics and culture. Since 1949, Beijing has become the one and only capital of the People's Republic of China.

In 2012, Beijing had twenty million inhabitants and occupied a territory as large as 16,808 square kilometers–that is ten times the size of Greater London. The city is organized in a series of six concentric rings. The first ring is synonymous with the outer walls of the "Forbidden City," the former emperors' palace. In the vicinity of the palace, the city also includes a number of famous individual buildings such as the Great Hall of the People and the National Museum of China. They are all located around the world's largest public square, Tiananmen Square. Five ring roads surround the historic center and individual city centers have been developing alongside.

Besides Beijing's historical and cultural importance, the Chinese capital is also gaining significance as a center for research and industry.

Beijing in Figures
Area: 16,808 square kilometers,
Population: 20.69 million

开篇感言

吕章申
中国国家博物馆馆长

中国国家博物馆新馆建成，对公众开放已两年有余。两年多来接待观众近千万人次，观众中有党和国家领导人，有外国国家元首、首脑和政要，还有国内外各行各业的专家、学者等。中国国家博物馆新馆建筑得到广大观众的广泛好评，特别是得到党和国家领导人、外国政要和建筑界专家等的高度评价。大家都认为国家博物馆改扩建工程是一个成功的例子，新馆与老馆建筑结合得很成功，新馆的室内外空间、功能、形式设计都很成功。得到这样的褒奖非常不易，作为业主方我感到十分荣幸。

中国国家博物馆新馆建设的成功，首先得益于设计出了一个好的建筑设计方案，这个建筑设计方案是由德国冯·格康，玛格及合伙人建筑师事务所(gmp)和中国建筑科学研究院(CABR)设计联合体设计的。这是中国国家博物馆新馆建设成功最重要的一个基本元素。所以我要特别感谢设计方所做出的杰出贡献！

从中国国家博物馆建筑设计的成功中我得到三点启示，讲出来与大家分享：一是在建筑设计方案招标中，得到的中标方案业主方感到不太理想或不满意怎么办？结论是一定要进行方案修改和完善，直到产生出一个大家都满意的建筑设计方案。但这需要业主方与中标设计方达成高度的共识。这说起来容易，但真正做起来难！这方面我们的设计方和业主方做到了，而且配合得十分默契。二是设计方一定要有一个愿意修改和完善方案的积极态度，并拿出最骨干的力量投入其中。在这一点上设计方做到了！让我十分感动！设计方团队无论是德方的胥茨先生还是中方的马立东先生，他们代表了设计方团队，表现出了高度的敬业精神和高超的设计水平。用最短的时间，修改出了大家满意的方案，这是十分关键的一环。三是业主方对建筑设计要懂行，否则业主方提出的修改意见会让设计方在修改方案时无法实施。这一点也非常重要，不然就会走弯路。

我十分庆幸：业主方遇到了德国冯·格康，玛格及合伙人建筑师事务所(gmp)和中国建筑科学研究院(CABR)这个十分优秀的设计团队，他们高度的敬业精神、虚心的工作态度和高超的设计水平成就了中国国家博物馆新馆建筑的巨大成功！也创造了一个改扩建工程建设成功的范例！

Opening Remarks

Lv Zhangshen—Museum Director of the
Chinese National Museum, Beijing

Since its completion over two years ago, the Chinese National Museum has been a magnet for large numbers of visitors. Well over ten million visitors have entered the museum and filled its galleries so far. This impressive number includes people from all over China, international tourists, as well as politicians and diplomats from China and abroad. The museum also exerts an enormous fascination on scientists and experts from a variety of fields and nationalities. As a result, we have received a great deal of public support and praise regarding the architectural redesign and extension of the Chinese National Museum. The international feedback from visitors, experts, and the media has been entirely positive. In particular, the combination of historical elements with contemporary architecture has been regarded as very successful. There has been unanimous praise for the urban integration, as well as the museum's interior and exterior design. We are proud that the design and its implementation fulfill all aspects of our functional and aesthetic aims. Given the myriad requirements and expectations, it proved especially difficult to find a common consensus during the design process. However, in my role as the client's representative, I always received both architects' full support and appreciation. I have felt very honored by this.

The architectural design—and therefore the success of the refurbished Chinese National Museum—originated from the cooperation between gmp Architects from Germany and the China Academy of Building Research (CABR). This positive collaboration was an important reason for the tremendous response we have had from all over the world. I express my sincere thanks to the design team for their continued effort and superb teamwork.

Since the successful completion of the Chinese National Museum I have taken three essential elements to heart: Firstly, what can be done if the client is not completely happy with the winning architect's design? The only solution lies in achieving a common consensus as quickly as possible. Any solution of that kind is usually developed through a laborious and intense exchange between client and architect; however, this is easier said than done. Nevertheless, in the case of the Chinese National Museum we achieved this through an admirable degree of close cooperation.

Secondly, the design team needs to be highly motivated to improve the design continuously. Experienced and ambitious individuals play a very important role in this process. Both architectural practices succeeded in doing so. I am personally grateful to Stephan Schütz, partner at gmp architects, and Ma Lidong from the China Academy of Building Research. These two gentlemen contributed to the project's success with their highly professional and creative abilities. Within a short period of time, they succeeded in improving the design to such a degree that in the end it received broad support. This was an essential milestone on the road to success.

Thirdly, at all times a client needs to have some architectural knowledge and an understanding of how to turn design into built reality. It is vital that a client should not obstruct the architectural design due to his lack of understanding. This item is of greatest importance, as it will inevitably lead to enormous costs and delays.

In August 2013, gmp Architects, who have been the lead architects in designing the Chinese National Museum, installed an exhibition in the Museum itself which shows gmp's design for the museum and other international gmp projects. I would like to take this opportunity to personally congratulate gmp Architects on their successful work. It has been an honor for me that I, as the client representative, had the pleasure of working both with gmp and the China Academy of Building Research. Thanks to their inspiring cooperation, both teams managed to successfully complete the design of the Chinese National Museum. It has become a widely recognized symbol of the refurbishment of a prestigious national monument.

转变与共识的印证
——记中国国家博物馆

马丁·罗特
伦敦维多利亚和阿尔博特博物馆馆长

时代的前行仿佛开足了马力。数年之后，当我们再回顾走过的历程，不免为眼前发生的近乎翻天覆地的变迁而触动。从今天的角度看，长驻中国从事大型建筑项目或赞助文化事业已不再是多么异乎寻常的举动，而其实不过是投身一场或多或少直面异域的对话。

这些今天司空见惯的事情，在15年前都是不可想象的。时代和大环境经历了飞速的转变，中国开放了，我们都是这一历史进程的一分子，而我们这里的经济发展已同中国难舍难分。

前瞻性的建设工程，譬如中国国家博物馆的扩建，不仅是开放的一部分，更是开放的催化剂与加速器。这座位于北京的博物馆是转变与共识的印证，它一方面为中国的民族意识与自豪感代言，另一方面象征着中国的开放。它彷佛一个矛盾，把政治的圆化为建筑的方，从不同的角度观察，它就会呈现出不同的面貌。在我眼里，冯·格康，玛格及合伙人建筑师事务所(gmp)的建筑师们都是代表自身专长的杰出外交家。他们在迥异的工作价值观之间架起沟通的桥梁，开拓了中德两国人的见识。感谢gmp建筑事务所和其他建筑事务所，他们同他们的国际团队不仅为中国建筑增添了新的维度，更使这一维度的引进成为普遍的共识。

毫无疑问，中国国家博物馆的扩建打破了一切现有的尺度，欧洲的建筑无论如何也难以望其项背。我曾有幸从2000年起就长期跟踪这个项目。尽管旧馆收藏着不计其数的奇珍异宝，但建筑尘封已久，亟待修缮。从旧馆到焕然一新、璀璨夺目的新馆，其间只经过了寥寥数年。无论根据怎样的国际标准加以评判，眼前崛起的这座博物馆都是无可挑剔的。

委托德国冯·格康，玛格及合伙人建筑师事务所(gmp)同中国建筑科学研究院(CABR)合作设计、规划、建造中国国家博物馆的项目显示出，中国早已不再通过闭关自守寻找新的自我认同，这一过程由建筑师们代劳了。

The National Museum as a Symbol of Change and Reconciliation

Martin Roth–Director of the Victoria & Albert Museum in London

When times change at such a speed that processes are viewed as history only a few years after they have occurred, one becomes aware that radical changes have taken place. From today's perspective, it doesn't seem very extraordinary to be working in China, to be completing a substantial number of construction projects, and to engage in cultural exchanges–in other words, to embark on a more or less direct discourse.

What today appears as an everyday scenario was unthinkable fifteen years ago. The times – and hence the conditions – have changed very rapidly; China has opened up and we are part of the opening process, not least because our economy is closely interwoven with that of China.

Pioneering efforts such as the conversion and new construction of the National Museum in Beijing are not only part of this opening process, but a catalyst and accelerator of it. The National Museum is a symbol of change and reconciliation, a sign of national awareness and pride on the one hand and of opening up on the other– a dichotomy that could be compared to squaring the political circle in an architectural context, which is perceived in different ways, depending on people's viewpoint. From my perspective, the architects von Gerkan, Marg and Partners are expert diplomats in their field, translators of very diverse work ethics and capable of expanding consciousness in China and Germany. It is thanks to gmp, and other practices and their international teams, that new dimensions have been introduced to Chinese architecture, and that everyday awareness has been shifted.

There is no doubt that the Chinese National Museum is off the scale – in every respect: the dimensions of the building are of a size that cannot be imagined in a European setting. I had the opportunity to watch the design process from the year 2000, and to follow it firsthand for a long time. Only a few years have passed between the time when the museum was a dusty place in need of an overhaul – full of copies and curiosities – and the current National Museum in its splendid new form. A museum has been created that compares to any standard worldwide.

The fact that the architects von Gerkan, Marg and Partners (gmp) and their local design partners, the China Academy for Building Research (CABR) were commissioned to design, plan, and build the Chinese National Museum, indicates that the search for a new Chinese identity is no longer a a process behind closed national doors, but one that is taking place on the world stage. The architects von Gerkan, Marg and Partners are making their contribution to this process.

序言

曼哈德·冯·格康

1998年，我收到北京德国学校项目设计竞赛的邀请，首次造访中国。

我参观了紫禁城、长城，它们都给我了留下了深远难忘的印象。当时我和我的同事就德国学校落成的时间打赌。后来我们在竞标中获胜，工程以惊人的速度完成，我也因此输掉了赌约。

中国的潜力是我当时丝毫没有意识到的。初到中国时，我虽然不怀任何偏见，但还是不得不承认，中国和我印象中的发展中国家形象相去甚远，这里几乎可以买到任何东西，路上有很多汽车，自行车却没有我想象中的多，酒店是欧洲标准，很多餐厅提供最高品质的服务。随处可见的财富和奢侈令人眼花缭乱。中国比我想象中走得更远。首次中国之行，我完全不曾料想中国会成为建筑和城市规划领域的潜在市场。

决定性的转折点出现在两年后，而且完全是一个巧合。1999年的国际建筑师大会(UIA)在北京召开，我应邀在会议上发表演讲，同时在北京一座名声显赫的博物馆里举办了我们的建筑作品展览。展览大获成功，媒体反响不凡，这也和组织方与政界、文化界重要人物的良好关系密不可分。这次展览将gmp推到了公众面前，为我们赢得了良好的声誉，树立了成功、可靠的当代建筑师形象。

长期以来，美国中西部的建筑形式外加装饰性的中国传统元素主导了中国建筑的形象，很多建筑最终呈现出漫画般的怪象。

gmp始终坚持自己清晰、简约的立场：材质的简洁、结构秩序的突出、功能的清晰显白，同时利用大自然给予的条件和特性降低能耗、优化采光。面对设计，我们首先期待可以借此占有最好的外部条件。从12年前直至今日，在几乎主导一切的"中国国情"下，我们的建筑设计理念已起到了某种示范作用，令我们在同众多美国同行的竞争中脱颖而出，也为我们赢取了200多项竞标和大量的设计委托。

在对永恒之美的追求下，唯有综合考虑客观条件、社会语境、功能需求、城市肌理（如果尚存的话）、经济发展等因素的建筑才是可持续的和历久弥新的。

就中国国家博物馆项目而言，我们的工作重点是通过我们纯净、简洁的形态语汇在对比之中化用中国传统文化的元素。有人云，并非模仿和适应，而是迥异的事物产生的张力才能促进建筑文化的进步——这一观点支撑着我们的设计理念。让新旧元素浑然一体，是我们首要的工作目标。除了文化和语言相去甚远之外，欧洲人与中国人在精神、气质和沟通、合作的方式上也大相径庭。我们西欧人，尤其我们德国人，不得不从中国项目中学会耐心的本领，即不能总是在明确目标后就迫使人尽快做出决定。我们还认识到，礼貌的中国人从不会斩钉截铁地说"不"，而"也许"这个字眼在某些情境下就足以表达"不"的意思。此外，丢脸对中国人来说是莫大的羞辱，所以，无论我们的谈判对手还是我们自己，总是不得不在言行中竭力避免让人失面子的事件发生。我们太过习惯直截了当、从不拐弯抹角的说话方式，如此情况下只好入乡随俗。

Preface

Meinhard von Gerkan

I had my first encounter with China when, in 1998, our architectural practice, gmp, was invited to take part in the competition for the new construction of the German School in Beijing.

The visit to the Forbidden City and the excursion to the Great Wall of China—these left a lasting impression on me. We took bets between colleagues as to when the school would be completed. We won the competition, and the building was completed in an extremely short time—and I lost my bet.

At that time, I completely failed to appreciate China's potential. When I traveled to China I had to recognize that my idea of this country as a developing country was rather mistaken: you could buy nearly everything, there were many more cars and less bicycles on the roads than I had anticipated, the hotels offered European standards, and some of the restaurants were of an excellent quality. There was much ostentatious wealth and luxury. The country was much more developed than I had anticipated. That China would become a future market for architecture and urban design was a thought that, at the time of my first trip to China, was completely absent.

The important change, barely two years later, was the result of an accidental event. In the summer of 1999, Beijing hosted the UI A Congress—the world meeting for architects. I had been invited to give a lecture and we were given the opportunity to stage an exhibition of our architecture in a well-known museum. The exhibition was a great success, also in the media, not least due to the host's excellent contacts with important personalities in politics and culture. Up to that date, there had never been an architectural exhibition in a museum in Beijing. There was something of an explosion in terms of public relations. As a result, we were able to build up a reputation as successful and reliable architects.

For a long time, developments in China had been dominated by American architecture combined with decorative elements from the Chinese tradition, which often ended up as a caricature of itself.

We were unfazed and insisted on our own position: rational, simple, reduced choice of materials, with an emphasis on the construction and functional logic, with natural properties, energy-saving and optimum daylight conditions. Initially we only expected to be afforded the position of a outsider, at best. Today, after twelve years, the "chinoise Sachlichkeit" (Chinese rational style)—which is close to dominating Chinese architectural style—has elevated our architectural approach to an exemplary level and, in well over 200 competitions, has earned us a substantial volume of commissions, also in competition with American colleagues.

What will ultimately prevail is an architecture that is generated sustainably and durably based on the actual conditions, the social context, the functional requirements, an understanding of the urban context (where this is still available), on the commercial requirements, and finally also on a timeless design intent.

At the National Museum, the transformation of traditional elements of Chinese culture and the contrasting combination with our purist/simple pattern language was the most important part of our work. Nevertheless, our belief that progress in architecture evolves from contrasting tension rather than from imitation and adaptation was confirmed. Our most important objective was to achieve a balance in the symbiosis of the old and new. Obviously, there are enormous differences between Europeans and the Chinese, not only in terms of culture and language, but also with respect to mentality and the method of communication and cooperation. Western Europeans, particularly we Germans, have to learn to be patient when working in China and to not be too focused on pushing for quick decisions. We have learned that a polite Chinese will never reply with a definite "no", and we have understood that sometimes the word "perhaps" may already mean as much as "no". We have also learned that loss of face is a serious issue for any Chinese and that therefore one has to strive to avoid this at all costs, both for oneself as much as for one's partner in the negotiation. For us, who are used to offering our opinions in a very straightforward and direct way, this required a fairly steep learning curve.

中国人仪式化的生活让我感到十分惬意。他们总是抽出时间，挽留匆匆而过的世间万物。中国人总是在商务洽谈或合约谈判的间隙安排社交活动，譬如共同进餐、茶话会之类。我们尊重这些活动，它们使人与人的沟通和相处变得愉快。处于放松状态下的交谈往往妙喻横生，而带有隐喻的谈话总是具有悦人的品位，尽管在某种程度上也显得模棱两可。始终让我们感到难以领悟和掌握的是中国人驾轻就熟的谈判技巧，每每这时，我们总觉得自己输人一等。

中国人充满活力，办事雷厉风行，对新事物抱有开放、好奇、勇于尝试的态度。他们不乏创新精神，并且能够迅速决策，从而把创意付诸实践。以上这些都给我留下了深刻的印象。某些事务在德国只能被慢吞吞地处理，在中国却常常得到麻利的解决。

在跨文化的对话中化冲突于无形

对话式设计的过程总是充满坎坷。广泛而持久的分歧和相互的说服工作常常伴随整个设计阶段，尤其当不同的文化背景导致甲方与建筑师观点相左的时候。

中国国家博物馆位于天安门广场侧旁，紧挨着紫禁城，中国五千年文化史的大部分都被浓缩在这一座毛泽东时代建造的建筑中。鉴于其日趋朽坏的构造、难以吸引公众的阴郁风格，旧馆亟需一次彻底的修缮。为此，若干国际知名的建筑师被邀请参加竞标。

针对毛泽东时代的旧馆，我们的竞标方案主张在一个体量庞大的顶层空间内安置占全部展览面积30%的巨大展厅。该展厅悬于"大广场"上方，可通过一段很长的电梯到达。我们设计的屋面被抛光的铜板包覆，并带有出挑很远的尾面。该设计的特别之处在于，来宾们可以在这个被抬高的展厅楼层内眺望天安门广场和紫禁城，以及其他聚集在这一带的中国最重要的国家级建筑。

设计伊始，合作还算和睦。方案越是深化，反对我们的声音就越发响亮，我们的方案被评为过于现代，我们设计的建筑（尤其新馆顶层的形态）被指为缺乏与天安门广场历史建筑群的统一与和谐。在这样的讨论中，一场旷日持久、错综复杂、各抒己见的对话拉开了帷幕。其结果是项目被叫停，除非我们准备接受中方的意见，努力确保檐口形态的连续性，项目才会继续。因为檐口构造在中国有深厚的传统，的确是中国最简洁有力的建筑构造形式之一。

经过我们内部漫长的讨论，终于得到了一系列修改方案，最终扬弃了顶层展厅，檐口的形态也改为逐层退后的出挑，从而呈现与周边的建筑和谐的面貌。同时，我们也得以在博物馆内部塑造出简洁、硬朗、现代感十足且气象宏阔的空间序列。

中国国家博物馆目前已经对外开放。它是旧与新、中国传统与国际现代主义建筑相互妥协的产物，后者谦恭地融入了当地的文化。

What I also find very appealing is a kind of ritualization of life. Not everything has to happen at top speed; instead, people take their time. We have therefore learned to appreciate that business meetings and contract negotiations often take place in the context of social activities such as a shared meal and toast speeches, which makes for pleasant human exchange and togetherness. Conversations are often cloaked in metaphors. Such metaphoric exchanges take place at a very agreeable level, although they may in some way be somewhat ambiguous. What we still struggle with, however, is understanding and mastering the clever methods of negotiating. Time and again, we feel less capable in this respect.

I am very impressed with the dynamism, the speed and, in a certain sense, the openness towards new approaches, the desire to try something new and to implement it—and that combined with really fast decision processes. What in Germany progresses at a snail's pace takes place at breathtaking speed in China.

Conflict Resolution via Inter-cultural Dialogue

› The design process involving dialogue can be a stony path. It is often associated with very extensive and persistent differences of opinion and the need to persuade each other, especially in situations where differences occur between client and designer due to cultural differences.

China has collected many artifacts relating to its 5,000 years of cultural history in the large museum, which was built during Mao's time at Tiananmen Square, directly adjacent to the Forbidden City. Because of its poor structural condition and its very dark and – for the public – rather uninviting character, this building was to make way for a new building for which an international competition had been organized among a limited number of well-known architects.

In contrast to the building dating from the Mao era, our competition entry included a large exhibition space—30% of the total exhibition capacity—in a voluminous attic floor, which was situated above a Grand Hall and could be reached via long escalators. We had planned to clad the roof with polished copper and include a far-protruding parapet. The special feature of this design was that visitors who ventured up to this large, elevated exhibition floor were offered a view of both Tiananmen Square and of the Forbidden City, as well as the vicinity's other very important representational buildings.

Initially, the design process proceeded in relative agreement with the state officials in charge of the project. However, as our design proceeded to take more detailed shape, more and more opinions could be heard that felt that our concept was too contemporary. The view was that the building planned by us was detrimental to the unity and harmony of the historic building ensemble around Tiananmen Square, particularly the design of the roof. This criticism led to a long-drawn-out, complicated dialogue with many voices, with the result that we would be suspended from further work on the project unless we were prepared to take into account critical comments from the Chinese side and make an effort to retain, above all, the uniformity of the roof parapets, which were considered to be part of a long Chinese tradition and which, in fact, are one of the most conspicuous architectural elements in China.

In turn this led—after long discussions within the practice—to a series of modified designs that finally omitted the exhibition floor in the roof and were externally more adapted to the other buildings, particularly with respect to the design of the roof parapet with staggered, receding projections. In turn, we were allowed to design the interior of the museum in a very contemporary, austere, and straight-lined style, which, most importantly, included a generous sequence of spaces.

The museum, which has now been opened, is the result of a successful compromise between old and new, between traditional Chinese and modern international architecture, with the latter deferentially fitting in with the local culture.

一次跨文化的握手
——记世界最大博物馆的建造

安德拉斯·桑托

博物馆是一个国家成就与雄心的不朽勋章。从卡塔尔到巴西，再到印度，全球所有新兴的国家都在建设新的艺术场馆，并且往往延请国际知名的建筑师主持。没有哪个国家能够比中国更加明显地体现出这一潮流。在中国，一年之内兴建百座博物馆是国家政策——照此趋势，十年之后就会涌现出一千座新博物馆。这个数目令人咋舌。几乎每个省份、每座城市都有正在建造的艺术场馆项目。这些建筑多半遵循粗放发展的模式，是当前中国对公共基础设施建设和软实力提升的划时代投资的一部分。

中国国家博物馆新馆坐落在北京市的心脏地带。它彷佛一艘满载各种象征符号并且再也装载不下更多的船。毕竟，它是中国最煊赫的博物馆，是一个拥有近14亿人口的泱泱大国对其人民和整个世界讲述自己的故事的场所。2011年，德国冯·格康，玛格及合伙人建筑师事务所(gmp)与中国建筑科学研究院(CABR)合作完成了新馆的设计。这座博物馆把中华文明的五千年历史浓缩在玻璃展柜之中，又另设展览向中国人民介绍世界文化。这是一座国际性的博物馆，它属于正把触角伸向五洲四洋的中国，因此它的设计必须体现出这一点。

为了实现上述抱负，中国国家博物馆的设计试图结合不同的主题。对作为甲方的中国政府而言，最重要的是建筑设计的表达必须同中国的历史文化明确相关。国博所在的基地位置要求其建筑必须与周边文脉协调一致。根据决定，中国国家博物馆将进驻一座建于建国初期的老建筑，后者本身就是一座久经考验的博物馆建筑。甲方相信，对它施行适当的改造将确保它能够同周边的环境产生有机的互动，而这样的互动会比大部分近年由西方建筑师设计的北京建筑更和谐。那些被视为具有纪念性和创意的建筑物大多与它们所处的城市环境格格不入。

国博新馆不仅讲述中国的掌故，也宣扬中国在世界上的地位。作为对这一诉求的回应，建筑师们小心翼翼地在中式设计与西式设计之间寻找平衡。他们把质朴节制的国际现代主义风格同频繁出现的中国传统文化元素连结起来。设计明白无误地传达出一个信号，即正在发展中的中国自我叙事再次同外部世界相连。建筑形态的转变为这座象征中国的博物馆绘制出一幅非凡的关于国家身份认同和自我形象的新地图。

俯瞰天安门广场（图1）这一中国历史、文化和政治的支点，从毛主席纪念堂的石阶和紫禁城的大门望去，中国国家博物馆的巨柱门廊便浮现在眼前。这座独特的建筑在1959年建国十周年庆典前夕落成，成为新中国"十大建筑"之一。国博前身以前苏联设计风格为蓝本，由清华大学和北京建筑设计研究院共同设计，张开济先生主持施工图设计工作，工程建设仅花

1

A Cultural Handshake: Building the World's Largest Museum

An Architectural Criticism by András Szántó

Museums are enduring emblems of a country's achievements and aspirations. From Qatar to Brazil to India, ascendant nations everywhere are building new art institutions, often with international architects. Nowhere is the trend more apparent than in China, where state policy has mandated the construction of one hundred museums a year – a staggering one thousand new institutions over a decade. Each province and city, it seems, has a project in the works. The buildings, many of them anchoring vast development schemes, are part of an epochal cycle of investment into public infrastructure and soft power.

The design of the new National Museum of China, situated in the heart of Beijing, is freighted with more symbolism than most. This, after all, is China's ranking museum – the place where a country of 1.4 billion tells its story to its citizens and to the world. Completed in 2011 by the German architectural firm of von Gerkan, Marg und Partner (gmp), in collaboration with local design partner CABR China Academy of Building Research, the museum showcases 5,000 years of Chinese civilization, along with exhibitions that introduce Chinese citizens to facets of international culture. It is a global museum for a globally connected China — and the architecture needed to reflect that.

The National Museum's design invokes a multitude of bridging and linking themes to serve these aspirations. Most important for the commissioning authorities, the building had to express a clear connection to Chinese culture and history. The site underscored the desire for a contextual approach. The National Museum, it was decided, would inhabit a historic structure dating back to the foundational period of the People's Republic– a building that already had a well-established legacy as a museum. Adaptive reuse, the Chinese clients believed, would ensure a more organic dialogue with the surroundings than is evidenced in many recent projects in Beijing by Western architects. Iconic and innovative as those buildings are, they appear to turn their backs, by and large, on their host environment.

The new museum would have to say something not only about China, but also about China's place in the world. The architects responded with a carefully calibrated balance of Chinese and Western design attitudes. They combined the understated vocabulary of international architectural modernism with ever-present allusions to Chinese cultural traditions. The design sends an unmistakable signal that China's evolutionary narrative is now once again tied to the outside world. The building's transformation has mapped onto China's emblematic museum the extraordinary realignment in the country's identity and self-image.

› Overlooking Tiananmen Square,[1] the historic, cultural, and political fulcrum of China, steps from Chairman Mao's mausoleum and the gate to the Forbidden City, the columned edifice of the National Museum originated as one of the Shi Da Jianzhu, or "Ten Great Buildings," hastily erected in 1959 to mark the tenth anniversary of the People's Republic. Construction of the original structure, based on Soviet classicist plans by Tsinghua University and the Beijing Architectural Design Institute (BIAD), with Zhang Kaiji in the role of lead architect, took just ten months. Through much of its life, the structure housed not one but two museums, one dedicated to Chinese history, the other to the Chinese revolution. The precursor to the former was the Preparatory Office of the National History Museum, created in the summer of 1912 – the first museum operated by the Chinese government. It was physically located, from 1912, at the Imperial College, and later moved to the Forbidden City. In 1949, it was renamed the Beijing National History Museum to mark the founding of the People's Republic; its name changed again in 1959, to the National Museum of Chinese History. The predecessor of the Museum of Chinese Revolution was the Preparatory Office of the National Revolution Museum. It was founded in March 1950 and given its current name

2

了十个月便竣工。这座建筑原本包含中国历史博物馆和中国革命博物馆两个部分。今日的中国国家博物馆的前身是成立于1912年的国立历史博物馆筹备处——这是中国政府创立并经营的首家博物馆。筹备处从1912年起位于国子监，后迁入故宫。1949年，在中华人民共和国成立的那一日，国立历史博物馆更名为国立北京历史博物馆。1959年，它又更名为中国历史博物馆。中国革命博物馆的前身是中央革命博物馆筹备处，最初成立于1950年3月，十年后更名为"中国革命博物馆"。1959年10月1日，一座南北长313米，包含中国历史博物馆和中国革命博物馆的建筑面向公众开放。随后的几十年间又经历过闭馆；直至2001年，场馆被认为其已不能够再体现中国日新月异的自我形象。2003年，两家博物馆合并为中国国家博物馆（图2），这一事件是国博漫长而复杂的历史中的一个转捩点。

年复一年地招待大批前来参观的学生和公务团体，老馆已经呈现出老态。馆中藏品，包括上百件来自全国各地的重要出土文物，已经无法在过时的展厅里妥善展出。随着2008年奥运会的临近，国家领导人决定扩建国家博物馆。该项目被列入国家十一五计划之中，并将在中共委员会和若干国家部委的监督下进行。为了表达出与时俱进的姿态，有关部门决定，扩建后的新馆不仅包括本土展品和艺术品的展陈空间，也设有能够举办全球巡回展览的厅室。同时，他们也热衷于把新馆建设成当代先进博物馆技术的集大成者。

这里存在着一个需要弥补的建筑不平衡。虽然老馆每一处细部的壮丽都不输于邻近的政府建筑，但因为1959年资金匮乏，与人民大会堂同在天安门广场东西轴线上的老馆没有具备能同宏伟的人民大会堂相均衡的体量。七年的修缮使国博的体量翻了三倍。当2011年3月1日国博重新对外开放时，国博被称为全世界最大的博物馆。

当然，体量是举足轻重的要素，因为它象征国家的自豪感。接触该项目的西方官员急需从中国同僚那里获得这座世界领先博物馆的精确尺寸。相比其魁梧块头的暗示，国博的建筑还传达出更多微妙的信息。项目的进程和成果均证实了建筑具有作为文化交流形态的强大功能。

当选择英国建筑师大卫·奇普菲尔德负责重建柏林博物馆岛上的新国家博物馆时，很多人都谈论过德国对集体记忆和集体历史责任的立场。同样地，选择一家德国事务所担纲中国国家博物馆新馆的设计，也意味着中国正在从新的角度反思自己。中国政府把如此敏感的项目交给国外的工程师负责，坦白了自己对西方专家技术的渴望。（毫不奇怪，国博新馆的首次国际交流展就是由德国博物馆联合会组织的，主题为欧洲启蒙时代艺术（图3）。）德国建筑师尝试了各种想得到的办法，把来自本土和国际的文化影响在他们的博物馆方案中交织起来，力图同时受到两种美学传统的欣赏，并且丝毫不失偏颇。他们的成果是一座由石头、玻璃和青铜制成的跨文化握手的巨大丰碑。

ten years later. August of 1959 saw the completion of the 313-meter-long building on the east side of Tiananmen Square, and on October 1 of that year, the two museums opened to the public. The ensuing decades were punctuated by periods when the building was closed; by 2001, the museum displays were no longer deemed reflective of China's changing identity. Prior to closing down for renovations, in 2003, the two entities were merged to form a new unified institution, the National Museum of China.[2]

Having been host to a stream of visiting students and official delegations over the years, the building was showing signs of age. The museum collections, including hundreds of architectural finds from excavations around China, were not served well in the outdated galleries. With the 2008 Olympics approaching, the leadership decided that the building was ready for a makeover. Commencing in 2004, the reconstruction project was included in the Eleventh Five-Year Plan, and it proceeded under close supervision from the CPC Central Committee and several key ministries. In a sign of evolving attitudes, the officials decided to make room in the expanded museum not only for Chinese objects and artworks, but also for traveling exhibitions from around the world. They were no less intent on making the museum a standard-bearer for advanced research and exhibition technologies.

There was a lingering architectural imbalance to remedy as well. Though every bit as imposing as the neighboring government institutions, the original museum could not be completed on a scale symmetrical with the colossal Great Hall of the People, across Tiananmen Square, owing to a shortage of funds back in 1959. The seven-year reconstruction would triple the museum in size. By the arrival of the March 1, 2011, reopening, the renovated building was heralded as the world's largest.

Clearly, for the Chinese authorities, size did matter. It was a token of national pride. Western officials involved in the project fielded urgent inquiries from their Chinese counterparts about the precise dimensions of some of the world's leading museums. Yet the National Museum building conveys a more nuanced message than its Herculean dimensions suggest. Both the process and the outcome of the project testify to the power of architecture to function as a form of cultural exchange.

› Just as the choice of the British architect David Chipperfield for the reconstruction of the Neues Museum, on Berlin's Museum Island, spoke volumes about German attitudes concerning collective memory and responsibility, so too did the choice of a German architectural firm for the National Museum suggest that China was thinking of itself in new ways. In handing over such a sensitive remit to architects from abroad, the Chinese authorities were admitting to a hunger for Western expertise. (It bears noting that the museum's first international exhibition was organized by a consortium of German museums and devoted to the European Enlightenment.[3])

For their part, the German architects went to great lengths to intertwine local and global influences in their vision for the National Museum. Their approach pays respect to both aesthetic sensibilities, without tipping too far in either direction. The result is a kind of giant cultural handshake set in stone, glass, and bronze.

3

包括雷姆·库哈斯和赫尔佐格&德·梅隆在内的著名建筑师也被邀请来为国博的扩建提出建议。按照项目大纲的说法，焕然一新的国家博物馆"应该相应地体现出中国充满活力的发展和人民日益增长的文化需求"。从技术的立场上说，这一大纲倡议复原旧时的布局，并给南北两个室外庭院加盖屋顶，即所谓的"留三面"方案。德国冯·格康，玛格及合伙人建筑师事务所(gmp)和中国建筑科学研究院(CABR)组成的设计联合体最终中标，淘汰了入围最后一轮竞标的另外两大竞争对手：诺曼·福斯特（同1959年设计老馆的北京市建筑设计研究院组成设计联合体）和美国KPF建筑事务所（与华东建筑设计研究院组成设计联合体）。

在博物馆的行业圈子里，总部位于汉堡的gmp建筑事务所的知名度虽然不及其他竞争对手，然而gmp建筑事务所曾担纲德国若干最大的基础设施项目的设计，其中包括柏林的老机场和新机场以及耗资十亿的柏林中央火车站。1999年，gmp建筑事务所登陆中国，在北京、上海和深圳设有分公司。它在中国的第一个项目是德国驻华大使馆附属的德国学校。多年以来，gmp建筑事务所在中国各地设计了众多体育场馆、物流中心和办公写字楼。它所中标的项目中还包括一座基督教堂、上海东方体育中心，以及重庆、青岛和天津市的大剧院。可容纳80万人口的临港新城总体规划和新城内一座帆船形状的航海博物馆亦是出自它的手笔。

国博扩建工程于2007年3月正式启动。gmp设计团队由曼哈德·冯·格康和斯特凡·胥茨领导，施蒂芬·瑞沃勒也加盟其中。他们从周边充满历史积淀的景观中获取设计灵感。从基地环望四周，故宫的琉璃瓦清晰可见，具有超大尺度和深厚历史意蕴的天安门广场（图4）也散发出强大的气场。建筑师的首要任务是把保存下来的既有建筑体同扩建的部分有机结合起来，创造出有说服力的内涵。他们打算在老馆的表皮之下安置一座现代化的艺术馆，并以此提升这一经典建筑的品质。

无论采用怎样的设计思路，gmp设计团队都力图使建筑更加贴近民众。今日的建筑师往往寻求把古典博物馆建筑开放化和人性化的方案，这么做一般会取得民众的敬意，却难以赢得他们的钟情。在gmp设计团队的竞标方案中，1959年老馆修长的立柱和透风的门廊为接下来进入体量急遽扩大的入口大厅提供了过渡。在建筑师的设想中，这间装潢一新的大厅简直就是天安门广场的延续（图5）。它位于北京的心脏地带，为陷于首都繁忙和恶劣天气的人们提供了恰当的避难所。假如把博物馆的内部空间设计成可遮风挡雨的公共空间，将会有怎样的效果？

4

Ten renowned architects, including Rem Koolhaas and Herzog & de Meuron, were invited to submit proposals. The refurbished institution, the guidelines stated, "should correspond with the dynamic development of China and the increasing cultural demand of the people." From a technical standpoint, the brief called for rebuilding the original structure and roofing over its two courtyards, the North and South Yards – the so-called Retaining Three Façades scheme. The architects von Gerkan, Marg und Partner (gmp) – CABR China Academy of Building Research team won the competition, beating two finalists: Norman Foster (allied with BIAD, which had also been responsible for the 1959 plans) and Kohn Pederson Fox (allied with ECADI East China Architectural Design & Research Institute).

Though less well known in museum circles than some of the other competing teams, the Hamburg-based gmp had been responsible for several of Germany's largest infrastructure projects, including Berlin's old and new airports and its billion-dollar central rail station. Present in China since 1999, the firm maintains offices in Beijing, Shanghai, and Shenzhen; its first project there was a school for the German embassy. Over the years, it has designed sports stadia, transport hubs, and office towers around China, and its winning commissions have included a Christian church; an aquatic center in Shanghai; theaters for Chongqing, Qingdao, and Tianjin; and a master plan for Lingang, a city of 800,00, complete with a maritime museum in the form of a sail ship.

The renovation was officially launched in March of 2007. The gmp team, headed by Meinhard von Gerkan and Stephan Schütz, together with Stephan Rewolle, took inspiration from the historic sites in the vicinity. The terra-cotta roofs of the Forbidden City are clearly visible from the museum. The scale and historic resonance of Tiananmen Square[4] were powerful influences. The task for the architects, above all, was to create a plausible cohesion between the remaining portions of the old structure and the newly built sections. They aimed to highlight the classic architecture while installing a thoroughly modern institution inside its envelope.

Above all, the gmp team wanted to make the museum more accessible. Today's architects often seek to open up and humanize older museum buildings, which typically command respect, but not so much affection from their audiences. In the submitted competition plans, the slender columns and breezy arcades of the 1959 structure formed a transition into a dramatically enlarged entry hall. The architects imagined this space as nothing less than an extension of Tiananmen Square.[5] The central core of Beijing offers scant refuge from the capital's bustle and inclement weather. What if the inside of the museum could function as a weather-protected public forum?

5

把博物馆塑造成大型公众活动舞台的理念也与近年来对博物馆定位的思考相符。从巴黎的卢浮宫，到伦敦的大英博物馆，再到纽约的大都会艺术博物馆，这些世界一流的艺术收藏机构纷纷展现亲民的姿态，宣称自己与市民生活的空间水乳交融。21世纪的博物馆不再只是专家和精英们向往的殿堂，它们立志成为公众的广场，安全可靠、全面开放，是不同背景的人群在购物、娱乐和礼拜的扰攘之余倾心交谈的地方。这些建筑设有的开阔广场、庭院、屋顶和花园也证实了博物馆正在转变为微型大都市的宏愿。

根据设想，巨大的入口大厅（图6）将迫使大部分展陈空间和附属空间移往上方楼层、地下室和建筑两翼。这一布局的关键是把常设陈列的展厅置于顶层。作为该设想的回应，gmp设计团队的第一套方案试图通过组织空间赋予建筑充满活力的形象：巨大的青铜屋顶（图7）盘旋在庭院上空，它以庞大的跨度包裹着一间蔓延上升的展厅（图8），从这里可以鸟瞰城市。在北京市最显赫的一些地点都可以看到常设陈列的展厅，包括天安门城楼和天坛。这样的解决方案不仅从功能上和形态上，也从来宾们的视觉感官上把博物馆和城市连结了起来。

悬浮的屋顶是国博新馆吸引来宾的巨大亮点。它象征着从老馆扬帆起航，而且与既有的巨型门廊不相连。后者同新造的内部空间泾渭分明，是扩建工程唯一不曾改动过的1959年老馆遗迹。对中国甲方而言，该方案过于激进。之后又出台了一些修正稿对方案进行重新评估。这些探讨都是在北京奥运会前夕发生的，当时中国建筑业界的观点逐渐趋向于反对采用来自西方的高技术方案。更有人质疑北京城越来越像先锋建筑的试验场。一些政府官员也对方案持有异议。国博的馆长吕章申，一位训练有素的建筑师，中止了方案的进程，把德国建筑师重新打发到绘图板前面。在短暂的休整并召集专家组开了几次会之后，馆长向建筑师提出在较低楼层增加展陈空间的需求，并要求博物馆的新旧两部分紧密结合为一个整体。

这是一个艰难的时刻。竞标方案的核心主张必须做出妥协。这个过程在巨大的压力下持续了一年。国博领导和建筑师三番五次地探讨方案，当地的设计伙伴中国建筑科学研究院也参与了协调，使德国的设计团队最终找到了解决办法。新方案的改动远远大于一次方案修改。他们调整了目标参数，重启了设计进程。在修正方案和最终方案中，屋顶只覆盖宴会厅、若干接待室以及一家餐厅；老馆的陈列移往较低楼层的展厅，底层的公共空间缩小一半，竣工日期也被推迟半年，即北京奥运会闭幕之后。这一过程的方方面面其实就是文化交流中的你来我往。我们今天见到的国家博物馆满载跨文化对话的烙印，是文化折衷的产物。

6

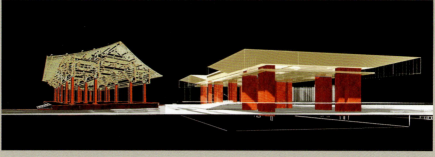

7

The idea of making the museum a staging ground for public activities dovetailed with recent thinking about museums. From the Louvre in Paris to the British Museum in London to New York's Metropolitan Museum of Art, collecting institutions are claiming a more populist role as our last truly civic environments. No longer just a refuge for experts and the elite, the twenty-first-century museum aspires to be a public square, safe and open to all – a place where people of all backgrounds can commune free from the interferences of shopping, entertainment, and worship. The expansive plazas, courtyards, roofs, and gardens in these institutions testify to the ambition to turn the museum into a microcosm of the metropolis.

As proposed by the architects, the gargantuan entry hall[6] would have required shifting the bulk of the exhibition space and various support functions to the upper levels, basement, and side wings of the building. The linchpin of this configuration would have been the repositioning of the permanent-collection galleries to the top floor. Correspondingly, the first gmp proposal was organized around a dramatic architectural gesture: an enormous bronze roof[7] that would hover over the courtyards, its vast span encompassing a series of exhibition spaces[8] overlooking the city below. The collection would thus have been presented in direct view of Beijing's key historic sites, including the Gate of Heavenly Peace and the Forbidden City. The floating roof would have been a powerful signature and a major audience draw for the museum. But it represented a departure from the existing architecture, including a marked disconnect between the existing façade colonnade – the last element from the 1959 building that would remain intact after the reconstruction – and the newly built interior volumes. The intervention proved too radical for the Chinese clients.

Several revisions ensued, leading to a fundamental reassessment of the plan. The discussions took place during the lead-up to the Olympics, when opinion in Beijing architectural circles began to turn against high-profile Western building projects. Skepticism was on the rise about the Chinese capital becoming a laboratory for experimental international architecture. Senior officials took issue with the plans, and the museum's director, Lu Zhangshen, himself a trained architect, halted the project and sent the architects back to the drawing board. After the convening of several expert panels, the director presented the architects with a need for more exhibition space on the lower floors and a demand for tighter integration of the old and new portions of the building.

It was a difficult moment. The central element of the competition design had to be scaled back, a full year into project development and under mounting time pressure. The museum leaders and the architects deliberated the matter with the local design partner, CABR. The German team concluded that the new guidelines amounted to more than a revision. They restarted the planning process with an entirely new set of target parameters. In the revised and final plans, the roof ended up housing a large multifunction hall (Jade Hall), a suite of reception rooms, and a restaurant. The historic exhibits were moved to lower gallery floors. The public space on the ground level was reduced by more than half of its originally proposed dimensions. The completion date was pushed back past the Olympics. All this, too, is part of the back-and-forth of cultural exchange. The National Museum we see today bears the imprint not only of cultural dialogue, but also of cultural compromise.

8

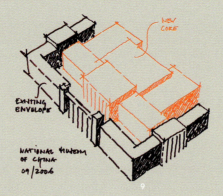

总建筑面积19.7万m²，内部空间可轻松容纳5万人，这座从旷日持久的规划工作中缓缓浮现出的造价3.8亿美元的博物馆毋庸置疑地展现了中国的宏图大业。它有48间展厅、一个712座的舞台剧场、一个264座的电影院、一间艺术大师级的数码演播室、科研图书馆、会议厅以及为来访的艺术家和科学家设置的酒店客房、宴会大厅和顶层观光餐厅——所有的这一切被建造出来，都是为了震撼人心。

从天安门广场望去，国家博物馆看来仿佛由三个大小大致相等的体块组成。较小的南翼楼和北翼楼被一道满载厚重装饰的柱廊连接起来。这一中央部分凸入广场之中，其额上是老馆留下的深红色的革命旗帜雕塑。门廊的方柱约40米高，它们身后是一条宽敞的露天过道，它把历史的立面与新造的建筑主体部分分隔开来（图9）。

国家博物馆的正立面在天安门广场的北部连绵了300多米，其侧立面还沿着车水马龙的长安东街（图10）延伸了200m远。修葺如新的旧立面从三面包裹一个崭新的内核。从远处看，唯一能提示出新内核存在的事物是增高的金属屋顶的轮廓；若从上方鸟瞰，德累斯顿雷瓦德景观建筑事务所设计的绿化屋面一览无遗。该事务所还为南庭院设计了抽象的中式水景，为北庭院设计了抽象的中式山景。

11-1

国博建筑最大的亮点不是它的屋顶，而是被恰如其分地称为"大广场"（图11）的那个空间——这是当初建筑师试图把天安门广场延伸至国博内部的设想的最后残迹了。虽然它比最初设想的小，却仍然惊世骇俗。它覆盖将近9 000m²的区域，比伦敦泰特现代美术馆的涡轮机大厅还要大25%。它是整个博物馆的接待中心，包含一系列诸如衣帽间、问讯台、咖啡厅、浴室和书店等来宾服务设施。（遗憾的是，国博新

馆开放以来增添的一些设施与空间的整体面貌并不协调。）从功能上说，大广场连结博物馆的南北两翼，是宾客人流的疏散中心。从这里开始，来宾分散前往各个空间，要么进入宏伟的社会主义、现实主义艺术画廊，要么从楼梯或电梯上楼进入内部展厅以及博物馆附属的教育、办公和餐饮场所。大厅的幕墙长260m、高27m，让整个空间沐浴在绚烂晚霞之中。这惊艳的入口大厅虽然已被缩小过，却仍不失为现代中国最有象征意义的内部空间之一。

一旦深入博物馆的内部，东方与西方的造型语汇彼此交融的细节令人目不暇接。虽然白色、四方的常规展厅占据了博物馆大部分的体量，然而对中式传统风格的提示俯拾即是。德国建筑师试图延续从1959年的设计就开始的一种跨文化的对话。当时的中国建筑师明显沉浸在欧洲的美学之中。他们创作出的建筑虽然也采用传统的风格元素，最明显的一处便是檐口优雅的陶土色镶边（一些镶边件在扩建期间被重新制作出来，并且仍由1959年的原厂家承担这一工作），却仍然具有明显的欧洲风格。整个20世纪50年代，类似的建筑风格也在东欧社会主义国家的首都方兴未艾。半个世纪之后，德国建筑师把对中国建筑原型的思考作为其工作的起点，然后从这一原型发展出新建筑的形态和材质。

10

› With a completed floor area of approximately 197,000 square meters and room inside to comfortably host 50,000 people, the 380-million-dollar museum that emerged from the planning process leaves no doubt about China's muscular ambitions. Its forty-eight exhibition halls, its 712-seat operatic theatre and 264-seat cinema, its state-of-the-art digital broadcast studio, its research library and conservation center, its hotel accommodations for visiting artists and scientists, its VIP reception rooms – all were built to impress.

Viewed from Tiananmen Square, the structure presents itself as a composite of three roughly equal masses. The slightly lower northern and southern flanks come together in a heavily ornamented portico. This central portion protrudes into the square, capped with a sculptural riot of crimson revolutionary flags preserved from the old construction. The rectangular columns of the portico reach up almost forty meters. Behind them lies an open corridor that marks off the historic façade from the new interior mass of the building.[9]

In addition to its more than 300 meters of frontage on Tiananmen Square, the museum stretches back 200 meters on Chang An East Road,[10] a heavily trafficked multilane thoroughfare. As per the competition brief, restored façades envelop the all-new central core on three sides. From a distance, the only clues to the modern inner core are the elevated bronze-colored rooflines. An aerial view of the building reveals two planted courtyards, designed by the Dresden-based Rehwaldt Landscape Architects, which created an abstract Chinese water landscape for the South Yard and mountain landscape for the North Yard.

The building's signature is no longer the roof, but the aptly named Grand Forum[11] – or what remains of the architect's desire to bring Tiananmen Square into the museum. Though smaller than originally envisioned, it is still breathtaking. Its almost 9,000 square-meter footprint is 25% larger than the London Tate Modern's Turbine Hall. This is the receiving hub of the museum, containing an array of visitor services such as cloakrooms, information counters, cafés, bathrooms, and bookshops.

Unfortunately, some features added since the opening do not conform to the building's overall design. Functionally speaking, the Grand Forum connects the northern and southern wings of the museum. It is a gathering point for visitors before they disperse inward through a large viewing gallery for socialist realist art, and upward, via stairs and escalators, toward the interior galleries and the educational, administrative, and hospitality spaces of the museum. Bathed in western light pouring in through a glass wall that spans the Forum's 260-meter length and 27-meter height, the staggering entry hall, even in its diminished form, is destined to be one of the iconic interiors of modern China.

Once inside the museum, the visitor cannot escape noticing the blend of Eastern and Western idioms. Although conventional white-box galleries take up much of the building, hints are everywhere to be found of Chinese styles and traditions. The German architects continue a conversation that started with the building's original plans. The original Chinese architects were clearly steeped in European aesthetics. They produced a structure that was ornamented in the traditional style, most prominently on its handsome terra-cotta-laced eaves (some pieces were remanufactured for the renovation in the same factory that had made them in 1959), yet was at once recognizably European. Similar buildings sprouted up in socialist capitals in Eastern Europe throughout the 1950s. A half-century later, the German architects began their task by imagining an archetype of a Chinese building. This would shape the forms and materials of the new structure.

11-2

于是，中国传统建筑的精华也已各就其位：整个建筑矗立在由当地花岗石筑成的地基上；内墙上的樱桃木板贴面暗示出中国古典建筑青睐木结构，它们首先被布置在一些具有礼仪功能的空间内；顶层礼堂内的樱桃木贴面外又覆盖了一层特别定制的吸音织物，而这层织物呈中国传统的深红色。（一楼的贵宾室由清华大学一位教授设计，她曾参与毛主席纪念堂和人民大会堂国家接待厅的设计。）檐口分层排布的白色金属梁低调地回应了周围建筑的传统飞檐。青铜的支架和稍稍隆起的窗玻璃都是来自中国传统五金工艺的灵感，却以减弱的西方艺术手法加以表现。一楼入口处宏伟的青铜大门（图12）带有镂空的花纹，可以把日光滤进大广场，给这个大教堂般的巨大空间增添了纹理和暖意。这些花纹也与馆中收藏的一座著名青铜鼎的花纹相似。

gmp设计团队在中国与西方的设计理念之间走出了一串令人叹服的优雅步伐。盲目的模仿将导致设计上的惨败，只会把国家的博物馆做成主题建筑。但凡尊重博物馆禁止陈列假古董的行业规范，刻意仿造的传统元素就必然不可取。另一方面，公然地让不同的风格彼此冲突则多半会得到某种眩目的后现代杂烩。所以，通过精心的协调和控制在两种传统之间做出微妙的取舍才是真正的挑战。

在材质的选择上，建筑师对自己提出了很高的要求，从而避免同传统设计产生过分直接的联系。譬如屋顶，其设计灵感虽然来源于紫禁城，却不采用传统的琉璃瓦铺设屋面，而是使用色调相似的金属板，给人更干净、更现代的视觉印象。他们把能联想到中国传统美的各色材质减少为三种——花岗石、樱桃木和青铜，辅以现代主义设计手法，创造出理性的统一感和建筑效果的一致性。在设计和建造的过程中，德国建筑师不得不克制自己的选择面。混用不同的材料、堆砌修饰的手法只会与中式的品味背道而驰。

对中国传统建筑的暗示不仅体现在建筑的结构和材质上。立柱、栏杆（图13）、大广场和其他豪华厅室的藻井（图14）都能让人联想到故宫。从入口大厅依次走过较小的展厅和更私密的空间的体验也同故宫的布局和流线设计相似。博物馆各处的设计都符合古老的风水学说。人们找不到一处锋锐的转角。流线设计上更多采用曲折的路径而非直接的路径。进入上方楼层的通道也不是西方博物馆中常见的集中式楼梯间，来宾们通过位于大广场两侧的一对楼梯上楼。这是一条迂回的路径，它让人想起风水学说中直接位于传统中国民居大门前用以挡住恶灵的障碍物——"鬼墙"。

此外尚有一处非常惹眼的中式元素，据说是无心插柳，大广场楼梯的石栏杆彷佛在模拟长城的轮廓。建筑师坚称这一处特别的关联只不过是巧合。或许德国的建筑师们已然深谙当地的建筑文脉，以至于在不经意间受到了启发？若真是如此，这将是证明文化渗透会给设计工作带来活力的最佳案例。

The essential components of traditional Chinese architecture are in place: The structure rests on a stone base made of locally sourced granite. Classic wood construction is indicated with extensive cherry-wood paneling on interior walls in the Grand Forum. The panels of the ceremonial hall on the top floor as well as on the entrance level and in the theater are clad in a custom-made sound-absorbing fabric in a shade of crimson – China's historic color. (The VIP rooms on the ground floor were designed by professor Zhang from China Central Academy of Fine Art who had once worked for the Mao mausoleum and on reception rooms in the People's Hall.) The staggered white metal beams of the new roofs faintly echo the traditional rooflines in the immediate vicinity. The large bronze entrance gates[12] on the ground floor are inspired by traditional Chinese metalwork, reimagined here in a pared-down Western aesthetic. They filter daylight into the Grand Forum through lacelike perforations that add texture and warmth to the cathedral- like immensity of space. They also form a visual link to a bronze vessel in the museum's collection.

The gmp architects had to walk a fine line when it came to finding a balance between Chinese and Western design philosophies. Blind imitation would fall flat, making the museum an exercise in theme architecture. (Ersatz copies, in any case, had to be avoided in deference to museum norms forbidding the use of artificial objects.) On the other hand, a too overt clash of styles might have created a jarring postmodernist pastiche. The challenge was to orchestrate a subtle give-and-take between two cultural traditions.

The architects imposed discipline on the choice of materials and avoided overly direct references. The roofs, for example, are clearly inspired by the Forbidden City, but instead of traditional ceramic tiles they are finished with colored metal for a cleaner, more contemporary look. Reducing the number of recurring materials to three – granite stone, cherry wood, bronze alloy – followed a modernist impulse to achieve unity and consistency throughout the building. The German architects had to press their case for self-restraint during the design process. It runs counter to the Chinese taste for mixing materials and embellishing rooms with all manner of decorations.

12

The allusions to Chinese building methods do not end with the building's structure and materials. Further references to the Forbidden City are to be found in the columns, balustrades,[13] and ceiling coffers[14] in the Grand Forum and other representative spaces. The experience of walking from large halls into progressively smaller and more intimate spaces echoes the structure and flow of the Forbidden City. Here and there, the building defers to ancient Feng Shui principles. Sharp corners and turns have been avoided. Indirect paths are preferred over direct ones. The upper level is not accessed through a central staircase, as would be the case in a Western museum. Instead, visitors ascend along a roundabout path, via a pair of stairs on either side of the Grand Forum. The gesture gives expression to the Feng Shui "ghost wall" ('ying bi'), a perpendicular obstruction placed directly in front of the entranceway of traditional Chinese houses, warding off intruding bad spirits.

One eye-catching reference was by all accounts unintentional: the stone staircase balustrades in the Grand Forum appear to mimic, in the abstract, the contours of the Great Wall of China. The architects insist that this particular reference was coincidental. Might the German designers have internalized the local context so deeply as to be inadvertently inspired by it? If so, this would arguably be the most genuine expression of the cultural osmosis fueling the design process.

13-1

跨文化交流不仅促进不同风格和传统的融合，也包含专家智慧的共享。就中国国家博物馆展览和科研的职能而言，这些影响主要是单向流动的，即从西至东。长期以来，中国博物馆一直处在西方的影响下。19世纪，欧洲兴建博物馆的时尚也被介绍到中国。作为集体认同化身的国家博物馆也是一个欧洲发明，并且在全世界得到狂热的仿效。虽然近年来中国各地的博物馆建设如火如荼，但它们更多来自行政命令，而不是真正的需要，设计上更多是对西方的效仿，而不是出于民族固有的性格。在中国，博物馆大体上仍是一种西方技术，一种足以让中国以令人印象深刻的决心和意志引进的技术。眼下，学者、博物馆馆长和科技人才为中国的博物馆发挥出的能量仍然逊于西方的同行，还有很多事物有待学习。

通过国家博物馆的扩建，中国政府已经创造出中国可经营的最先进的博物馆。这座博物馆将确立博物馆建设的新标准。尽管其体量过于庞大，但是其操作值得进一步精细化，从对其博物馆学、科研、教育等功能的经营，到建筑外表背后庞大的技术部署。而德国正是开展博物馆经营和技术方面知识交流的天然伙伴。西方之所以有其博物馆理念，首先要归功于德国。中国一向高度推崇德国在基础设施建设方面的专门智慧。两国专业人才间的交流促成了十多年以来中德在建筑领域的合作。在项目的进程中，德国建筑师们从数不清的角度斟酌过博物馆的设计，已经超出了原来的基本要求。他们建议更新照明、通风和温度调节系统，又建议铺设绿化屋顶以贮存雨水和调节室温。施工上的每一处细节都在他们的监理下执行。在一个缺乏精心规划的博物馆的国家里，扩建中国国家博物馆的任务在同德国建筑师持之以恒的互动下推进。欧洲、亚洲、非洲和美洲的博物馆数不胜数，在全球范围内委托博物馆的建设项目并将其推进到其最终的形态得益于持续的跨文化合作，此言非牵强之论。

从某种程度上说，中国以超大规模投资文化设施建设的驱动力正是在与世界的交流过程中获得的。这是一个值得注意的观点。全国各地正在兴建的数百座博物馆代表着国家获取国际声望并吸收海外文化影响的意愿，而这些文化影响正是中国公民需要的，他们可以借此理解这个世界，并将在这个世界中发出更有影响力的声音。中国目前的文化输出完全能与最大的工业国相提并论。根据报道，中国的文化输出每年增长15%；大部分新开展的文化活动具有国际背景；迄今已有800多亿美元被投入用于软实力的提升，比如举办2008年奥运会和2010年世博会。1996年以来，中国艺术市场的市场份额已经翻了两番，成为世界最大的艺术市场之一。数百家孔子学院和数千名艺术管理人员通过日复一日的辛勤工作促进着中国与其他国家之间的交流。

13-2

14

Cultural exchange involves not just the amalgamation of styles and traditions, but also the sharing of professional expertise. When it comes to the National Museum's exhibiting and research functions, the influences flowed mainly in one direction – from West to East. This has long been the case for Chinese museums. The European fashion of building museums was introduced to China in the nineteenth century. The notion of the national museum as an embodiment of collective identity is, of course, a European invention that has been enthusiastically copied worldwide. Although museums have proliferated in China of late, they remain, by and large, a Western technology – a technology that China is adopting with impressive resolve. For now, however, the scholarly, curatorial, and technical capabilities of Chinese museums still lag behind those of their Western counterparts.

For the refurbished National Museum, the authorities set out to create the most operationally advanced art institution in China. The museum would need to set a new standard not just with its size, but through the sophistication of its operations – from its museological, research, and educational functions to the vast array of behind the-scenes building systems. Germany was a natural partner for this exchange of technical know-how. Western notions of art museums owe as much to Germany as to any nation, and Chinese opinion has long held German infrastructural expertise in high regard. The professional relationships that led to the architectural collaboration with Germany stretched back more than a decade. Over the course of the project, the German architects weighed in on countless aspects of the museum, going well beyond their core design mandate. They proposed up-to-date lighting and acoustic solutions as well as ventilation and climate systems in the exhibition halls. They watched over the execution of every construction detail. In a country where museums often are conceived with scant advance thinking about their programming, the mission of the new National Museum evolved through sustained interaction with the Western architects. It is hard to say where such influences begin or end. Yet it is not too far-fetched to suggest that the broadening of the museum's global mandate—resulting in dedicated galleries for exhibits from Europe, Asia, Africa, and America – may have owed something to the prolonged intercultural collaboration that propelled the institution to its final form.

› A remarkable aspect of China's epic investments into cultural facilities has been the degree to which they are driven by an impulse to communicate with the world. The hundreds of museums now being built around the country are intended to project China's prestige abroad and to absorb cultural influences that China's citizens will need to operate in a global society in which they are claiming a larger role. China's cultural output now ranks with those of the largest industrialized nations, and is growing rapidly. Much of the new cultural activity is designed with the rest of the world in mind. More than eighty billion dollars were spent on huge soft-power initiatives, such as the 2008 Olympics and the 2010 World Expo. The Chinese art market has increased its market share more than four fold since 1996, becoming the second largest after the United States. Hundreds of Confucius Institutes and thousands of arts administrators labor each day to nurture engagement between China and other nations.

每年建造一百座博物馆的政策堪称全球亘古未有的博物馆建设热潮，这也是中国的全面开放政治中不可或缺的一部分。中国投资博物馆建设的成功并不能只用数据进行评价。真正的评判准绳是通过建造的这些场馆所学习到的经验的性质。该准绳亦适用于中国国家博物馆。关于它的统计数据，包括空前庞大的体量、120万件展品和800万来宾，都毫无疑问地令人敬畏。但是，它是否真正成功，取决于它是否能为中国各地的博物馆建设提供参考标准，而且更重要的因素是，它是否能培养出一个具有全球视野和自我意识的中国公民群体。对博物馆而言，这些期待远远超出了建筑本身，却是从建筑迈出第一步。

在未来数年中，中国的博物馆会继续努力在国际上站稳脚跟，同时还会继续寻求它们在中国国内文化生态中的地位。它们的工作人员编制和所举办的展览数量将跟随中国的经济一同增长。倘若中国最重要的博物馆为中国各地正在萌芽的博物馆建设上了一课，那么其中心思想就是文化不可能来自闭门造车，它以对话和互惠为生存养料，还需要倾听的技巧。文化是在过去与现在、内部和外部、我们和他们之间追求均衡与和谐。中国国家博物馆新馆告诉我们，虽然这一追求的终极目标不甚清晰，但只要人们不惧怕同他人分享自己的见识与才华并且联合起来，文化就一定能够繁荣昌盛。

原文发表于美媒《The Art Newspaper》
2011年4月刊，
略有增减修改。

The policy of erecting one hundred museums a year – most likely the largest museum-construction spree the world will ever witness – is part of this grand opening. Yet the success of China's museum investments should not be measured only in numbers. The most relevant outcome will be the quality of the learning experience that the new institutions provide, the subtle effects they are destined to have on individual identities and collective bonds. The same is true of the National Museum of China. Its vital statistics – its record-size building, its 1.2 million objects, its eight million visitors – are awesome beyond dispute. The true measure of the museum's achievement, nonetheless, will be whether it can set progressive norms for other museums around China, and more important, whether it can help nurture a more globally informed and self-aware Chinese citizenry. For museums, such aspirations admittedly go well beyond architecture. But they begin with it.

In the years to come, Chinese museums will continue to seek their footing in terms of global relevance and with in China's own cultural ecology. Their staffs and exhibitions should improve with time, benefitting from better training, more experience, and an ascendant economy. If the most important museum in China has a lesson for the many new institutions now sprouting up around the country, it is that culture cannot prosper alone. It feeds on dialogue and reciprocity. It demands listening skills. Culture consists of a perpetual search for symmetry and harmony – between past and present, inside and outside, us and them. The National Museum of China's building tells us that this search, although its ultimate destination is seldom clear, is best undertaken in alliance with people who are not afraid to share their visions and talents.

This essay was adapted and significantly expanded from an article by the author about the National Museum of China in "The Art Newspaper", April 2011.

记一次中德合作

对话——斯特凡·胥茨(gmp)和马立东(CABR)
主持人：周榕

周：很高兴能有这样的机会跟胥茨先生和马院长进行开诚布公的交流。自1998年进入中国以来，德国冯·格康，玛格及合伙人建筑师事务所(以下简称gmp)已经在中国大陆完成了150多个设计任务，其中不乏诸如国家博物馆这样重大的公共建筑项目，在业界具有很高的知名度和广泛的影响力。我们看到，在过去的十几年间，大量境外事务所涌入中国设计市场，其中"水土不服"的案例俯拾皆是。而gmp在中国却一路顺风顺水，取得了越来越引人瞩目的巨大成就，成为具有旺盛生命力和高度竞争力的标杆性国际事务所。请问胥茨先生，gmp在中国获得成功的秘诀是什么？

胥茨：不能说有什么秘诀。我想gmp成功的关键在于秉持一种基本态度——"对话式设计"，这个基本态度，自gmp事务所成立以来，就始终在每一个项目中贯彻执行。所谓对话，首先是指围绕建筑项目所进行的一切交流活动。其次，对话当然也指探讨、解读、阐释建筑项目所在地的文化。最后，我们的工作永远都是在与反映了业主与使用者特定需求的某项建造任务进行对话。只有建筑师准备好时刻面对这一对话，从复杂多面的设计任务中才能诞生出独一无二的建筑作品。对此，不论是来自建筑师个人的文化背景，还是从陌生环境中汲取的文化经验，都有着极其重要的意义。

周：我想就您刚才提到的三点展开进一步的探讨。众所周知，gmp在中国投标成功率很高，但德国与中国的文化传统以及思维方式之间存在着明显区隔，相对而言，中国文化复杂多变、偏于情感因素，而德国文化则推崇理性，思维逻辑严谨、操作技术精密。德国人理性的思维方式与中国人较为感性、灵活、强调形象意义的思维方式之间存在极大的差异。在中国殊异的文化语境中，gmp是如何准确把握业主意图的？又是如何以同时具有建筑品质并且业主能够理解接受的设计语言对业主意图进行诠释的？

胥茨：如果我们一旦了解中文本身就是一种纯图像式的语言，就完全可以理解形象和象征在中国文化中所具有的重要意义。中国一方面在经济上正以不可思议的速度向前发展，另一方面也能够站在历史的高度上回望传统。随处可见的一个问题便是，如何将二者结合起来？

一个正在传统与现代之间寻找平衡点的泱泱大国，其自我身份的认同感（identity）是基于图像或隐喻的。作为建筑师，我们一方面要认真地接受这一要求，另一方面也要根据情况的不同而区别对待。我认为，在诸如重大公共建筑的设计上，业主提出对象征性、强烈形式感的要求，是完全可以理解的。这类建筑代表着城市的形象，具有文化上的标志意义。然而，办公与住宅建筑则不需要过分强调象征性，它们实际上构成了城市的环境基底（context）。也就是说，城市的建筑项目，有着清晰的类型上的分别。如果不加以区分，我们的城市便会陷入一种建筑"噪音"(cacophony)的状态。很遗憾，在中国很多城市，这样的噪音状态已经初露端倪。

德国的情况与中国恰好相反。在德国，我们早已不再谈论城市扩张，而是谈论城市收缩。即使是首都柏林，与人口增长相比，城市发展也已近乎停滞。德国城市的文化特征，在很大程度上已经在城市发展过程中趋于固定，除了个别案例外，很少见到需要通过标志性建筑来创造城市认同感的倾向。总体而言，中德两国不同的经济、社会、政治条件决定了对建筑与城市的迥异需求。

A German-Chinese Cooperation

An Interview with Stephan Schütz(gmp) and Ma Lidong(CABR)
conducted by Zhou Rong

Zhou Rong   I am very pleased about the opportunity to be able to talk in detail with Stephan Schütz and Ma Lidong. Since 1998, when the first gmp office was opened in China, von Gerkan, Marg and Partners have completed over 150 projects in China. These include very important public projects such as the Chinese National Museum. Although numerous foreign practices have entered the Chinese architecture market over the last ten years, many of their projects simply do not fit in with the place at which they were built. That is different with gmp, and that is why the practice is so successful in China. gmp has become a very lively and competitive international practice. Stephan Schütz, are there any "secret tips" that might explain this success?

Stephan Schütz   I would not say that there are secret tips. But I believe that it is our fundamental approach to architecture which has been systematically developed since the opening of gmp's office. This approach shines through every project completed by gmp. It can be described by the term "dialogue" or "dialogue in design." The term "dialogue" refers to an exchange about what is going to be built and also an interaction with the culture and the spirit of the place of our projects. Ultimately our work always involves a dialogue, with the specific objective of reflecting the needs of both building owners and users. This dialogue is needed in order to design buildings that create something unique from the complexity and multiplicity of requirements. The experience from our own cultural background plays just as important a role as those elements we have to learn from the culture of foreign spheres.

Zhou Rong   In spite of the big differences between Chinese and German culture, gmp has been successful in many public architectural competitions in China. In many ways, Chinese culture is more flexible and variable. It is influenced much more by emotional factors, and symbols are often used as an expression. By contrast, German culture has a much more rational approach to dealing with the world. Germans think rather more logically and with responsibility, their method is very precise. The question is: how does gmp manage, in spite of these enormous cultural differences, to exactly capture the client's ideas and turn them into architectural reality?

Stephan Schütz   When we remember that the Chinese language is a picture language, it is not so difficult to understand the Chinese image- or symbol-based approach. China is a nation that has developed very quickly commercially and, at the same time, can look back over a very long tradition. The question that can be sensed everywhere is: how can both be combined?

The identity of a powerful country searching for a balance between tradition and the Moderne is also based on images or metaphors. As architects, we take this fact seriously while also striking a balance. I believe that the desire to build something with a powerful symbolic context is justified, particularly when it comes to important public buildings. They are an important element in the appearance of a city, and determine its image. However, I feel that metaphoric images are misplaced in offices and apartment blocks, which are more reflective of the urban context. This means that there is a clear typological difference between different building projects. Were this not so, our cities would drown in the multiplicity of architectural symbolism and metaphor. Unfortunately, that is the case in some places in China.

In Germany the conditions are different anyway. In most cases, we are not talking about growing, but about shrinking cities. Even here in Berlin, the city's development is stagnant in terms of the population size. To a large extent, the identity of our cities is predetermined by their history, and the tendency to create uniqueness through iconographic buildings is, exceptions excluded, relatively rare. In summary, we can say that different economic, social, and political conditions lead to different objectives for architecture and urban design.

周： gmp在中国有大量的建成作品，您认为其中有哪些堪称代表作？

胥茨： 由于每个项目的产生都有其特殊条件，每个项目背后都有一段特殊的故事，因此很难挑出某个具体项目作为我们最杰出的代表作品。但是在此我想重点以一个项目为例，这个项目可以说充分展现了我之前所谈到的"对话式设计"策略，它就是位于北京的中国国家博物馆。

众所周知，我们通过国际竞赛赢得了这个项目。当时的中标方案着重强调了老博物馆与新建部分之间的对比。我们设计了一个看似漂浮在老建筑上方的大屋顶，使其成为城市空间中一个极为明显的标志物，表现了国家博物馆诞生的各个历史时期。我们与当时的业主在不断的交流过程中一起对中标方案进行了长达一年的深化设计，直到有一天我们彻底改变了想法。业主认为这个方案没有给予天安门广场和老博物馆足够的尊重，因此没有为项目指明一个适合的发展方向。

最终的结果是，这个项目逐渐发展成为对老馆进行扩建，新建部分采用现存建筑的语言，用当代的手法进行抽象。我们拾取了老馆的建筑元素，加以变换运用到新馆中，令访客可以感知到建筑年代与阶段的不同以及建筑的诞生过程。坦率地讲，这个项目中的一系列对话过程曾令我们对自身的位置提出了根本性的质疑。我必须说，国博项目是我们整个公司历史中最极端和独一无二的例子——它证明了建什么以及怎么建的问题并不完全由建筑师决定。我们的城市与建筑都是一系列集合式程序的结果，而建筑师正是凭借个人化的、已有的建筑风格以及纯艺术化的手法在这一程序中登场，并对每个对话做出总结陈词。我想，对于所有成功的建筑作品，开放的设计态度不可或缺。

周： 建研院作为国内一个重要的、有着悠久历史的大型设计院，与gmp这样一个成熟的德国事务所在一起合作，双方不仅要面对中德两国之间巨大的文化差异，还要面对两种完全不同的设计机制、组织体系之间的差异，能否举例说明双方在合作过程中是如何对话沟通的？

马： 实际上，我们与gmp的合作到国博已经是第十六七个项目了。正如胥茨先生所说，国博既是gmp在中国最成功的项目，也是建研院至今最成功的项目。我认为成功取决于很重要的两点：一是gmp有着十分清晰的目标，胥茨先生已经表述过了；二是我们双方都秉持对话的原则。一般项目的主要对话方包括合作院、业主、施工企业。在国博项目中对话方又多了两个，一是专家方面，以吴良镛、周干峙、张锦秋、李道增、彭一刚、齐康、马国馨等诸位院士为代表的国内顶级专家团队；二是政府，例如发改委等机构。他们的介入以及对项目造成的影响，与其他项目相比分量要重得多。多达五方之间的对话，使得项目的推进和决策过程变得异常复杂。

下面我试举两个例子来说明这种对话的复杂性。

Zhou Rong   gmp has completed numerous projects in China. Which of gmp's projects in China do you consider to be the most successful, or most representative?

Stephan Schütz   It is very difficult to pick out individual projects, because each project has its specific conditions and specific history. Nevertheless, I would like to discuss a project that perfectly reflects the approach of "dialogue in design"– the Chinese National Museum in Beijing.

As you know, we won the project on the strength of a design for an architectural competition that focused heavily on the contrast between the existing building and the extension. A modern roof floating above the existing building was to create a landmark that can be seen from afar, making visible the different time epochs of the creation of the National Museum in the urban context. For an entire year, we refined our competition design in a dialogue with the then clients, until they changed their minds. The previous approach was considered to be inappropriate for the development of the project, because apparently it did not sufficiently honor the historic importance of the square and the existing museum.

As a result, this led to an extension building that seized on the pattern language of the existing building, but in a more abstract, contemporary way. Some elements of the existing building were adopted and transformed so that the history of the building's development becomes apparent for the visitor. This interactive process leads us to fully question our own position. I must add that this is a unique and most likely extreme example in the entire history of the practice. But it also bears witness to the understanding that we, as architects, are not the ones to decide what is built and how, and that our cities and buildings are the result of a collective process. In my view, if an architect applies his or her own individual predetermined style, he or she will be closed to this type of dialogue. I think successful architecture requires an open attitude.

Zhou Rong   The China Academy of Building Research (CABR) is one of the most important and traditional architectural institutes in China. The design cooperation between gmp and CABR was a considerable challenge. Both practices had to come to terms with the big differences in the cultural, design, and organizational thinking. Can you perhaps give a few examples of how the two practices communicated with each other during the time of their cooperation?

Ma Lidong   First of all, I must say that before the Chinese National Museum project, we had already worked with gmp on fifteen or sixteen projects. But to take this as an example: as already indicated by Stephan Schütz, the extension of the museum is probably gmp's most successful project in China. The same is true for CABR. In my opinion, this success is the result of two things. Firstly, gmp had a very clear objective – this has already been touched on by Stephan Schütz. Secondly, both practices consider communication to be very important. Nevertheless, in this case that was particularly complicated. In normal projects, the communication has to take place between the architects, the client, and the construction company. But in the case of the Chinese National Museum, two more parties were involved: the team of experts and the government team. The team of experts consisted of Wu Liangyong, Zhou Ganzhi, Zhang Jinqiu, Li Daozeng, Peng Yigang, Qi Kang, Ma Guoxin and others. Representatives of the government included the National Commission for Development and Reform and other authorities. These two additional teams played a decisive role and had a big impact on the project, even more so than in other comparable projects. So the fact that a total of five partners had to cooperate and agree with each other made the processes and decision-making extremely complicated throughout the entire design phase.

I will try to explain this using two examples.

第一是纯学术的：胥茨先生曾对老博物馆的风格定位提出质疑，它的风格到底是中国传统的还是西方的？当然也可以认为是中西合璧的。国博入口处有两个蒙古包样的体量，中间有一个柱廊，有西方古典建筑的味道。我们当时在这方面有过很多的探讨。之后我对他介绍了中国近现代建筑史的发展脉络，还有当时建筑师受到的苏联的影响，对中国传统的再发掘，接受西方教育后对西方建筑的崇尚，等等。当时在专家评审会上我斗胆地评说老国博属于折衷主义建筑，齐康先生立即站出来纠正说应该是古典主义。

诸如此类的讨论有很多，其实是否达成一致与形成结论都不重要，重要的是大家都积极地进行沟通并且促进了各方对项目本身的理解。

二是关于新馆的檐口。这个课题应该是探讨时间最长，反复最多的。中间做了很多1:1的样板。檐口部分一方面延续传统，另一方面也体现时代性，可以说是用现代技术反映传统理念。我们曾有过的争议主要是因为gmp一贯的设计理念是反对纯装饰的，比较讲究结构、材料、功能和建筑艺术的高度统一。

通过三方——吕馆长、建研院、gmp——反复的沟通、对话及验证，最后还是取得了一个相当成功的结果。这点胥茨先生可以进一步谈谈。

胥茨： 屋顶建筑语言决定了天安门广场旁与故宫内的所有建筑的形象。屋顶可以说是中国古建筑中高于一切的建筑元素。天安门广场上的所有建筑中，每个建筑师都曾试图用各自时代的语言来阐释故宫层叠的屋顶形式。对此他们大都采用了极为直接的方式，例如直接复制故宫建筑屋顶琉璃瓦的材料与颜色。与这种做法不同，我们则更希望在国家博物馆的扩建部分中清晰地表明，我们不想模仿古典的传统屋顶构造形式，而本来事实上新建部分的屋顶也是钢结构的，与故宫建筑的木结构屋顶有本质区别。此外，新馆层叠的屋顶构件中还依据功能、技术需要隐藏了大量的设备内容。屋顶材料采用了黄铜色的金属板与水平的金属百叶，最终呈现了一种技术化的文化形象。

周： 德国可以说是现代主义建筑最重要的发源地，而现代主义在中国却长期处于缺失的状态。gmp的设计哲学，是否继承了现代主义建筑的某些精神？ gmp又是如何在中国项目中表达现代主义精神的？

胥茨： 我对于现代建筑的理解首要是：目的与形式的一致性。因此我们在进行建筑设计时会尽量遵循来源于建筑任务目的的一切要求。我们希望通过将所运用的建筑手法尽可能简化，创造建筑美学上的独特品质。

The first example relates to the architectural style. Stephan Schütz had thought much about the architectural style of the old museum. What style does the old museum building represent? Is it a conventional Chinese style, or rather more a European one? Of course, one can also describe the building as a combination of Chinese and European styles. At the entrance of the museum there are two towers, the roofs of which resemble that of a Mongolian yurt. On the other hand, the design of the arcades between these two parts of the building evokes associations of a classic European style. We had lengthy discussions about that at the time. During these discussions, I told Stephan Schütz about the development of Chinese architectural history in the previous century: for example, about the influence of the Soviet Union on Chinese architects at that time, the search for Chinese traditions, and the trend of Chinese students to study in Western countries and then proceed to design architecture with a strong European influence. To the jury, I had referred to the National Museum as an eclectic building. But Kang Qi immediately corrected me: the old museum, he said, represented the classical style.

There were many debates of that kind. The point was not so much that of achieving a consensus, of drawing the same conclusions, etc. Rather, it was important to carry out intensive communication because the result was that all those involved were able to understand each other—and the project—much better.

The second example is about the roof projection for the new museum. This was the subject that was discussed the longest. Many detailed proposals by the architects were produced as 1:1 scale models and assessed. On the one hand, the roof projection was supposed to reflect and expand on the Chinese building tradition. At the same time, it was to appear modern and contemporary in all respects. The idea of the roof projection is to interpret tradition with modern methods. This issue was one of intense debate between gmp and us. We were under the impression that gmp had a problem with decorative elements, with ornamental patterns, because the practice considers a holistic design—in which construction, materials, function and aesthetics all interact—as very important.

After numerous discussions between the museum's Director, Lu Zhangshen, CABR, and gmp, we finally found common ground I think that Stephan Schütz may also be able to say a few words about that...

Stephan Schütz   The architecture of the roofs has a big influence on the appearance of all buildings in the Forbidden City and around Tiana-men Square. Generally speaking, the roof is the royal element of Chinese architecture, which is why the architects of all buildings at Tiananmen Square have tried to translate the staggered roofs of the Forbidden City into a contemporary form. They have done this in a very direct way, for example by simply adopting the same material of the Forbidden City's glazed roof tiles, including the color. By contrast, we wanted to make it clear in the extension to the National Museum that the roof construction was not a traditional one, if only for the fact that our roofs are made of steel and not of timber as in the Forbidden City. Furthermore, there is the fact that the staggered roofs of the new building accommodate numerous technical elements. This gave the roofs a technical appearance, with their bronze-colored metal plates and slat-like horizontal elements. We tried to bring together traditional and contemporary requirements in a new type of roof architecture.

Zhou Rong   Germany is considered to be one of the birthplaces of modern architecture. But the term "modern architecture" has been a foreign phrase in China for a very long time. Does the essence of modern architecture—however difficult that may be to define—play an important role in gmp's design philosophy?

Stephan Schütz   For me, modern architecture is primarily an interaction between purpose and form. For this reason, we design our architecture according to the detailed requirements that result from the overall purpose of the building. The aesthetic quality of our buildings develops from reducing stylistic means to those architectural elements required in the construction. However this cannot be a purpose in itself!

建筑设计的目标在于创造一个供人使用的、舒适宜人的空间。我们在所有的建筑项目中都试图将原本的设计任务书进行扩展，即基于设计任务的基础，尽可能创造出多样化的、宽敞的公共或半公共空间，为人们提供舒适的环境，人们可以在其中交流、活动。这一空间可以有各种各样的尺度与意义，可以是国博的中央大厅，可以是青岛大剧院、天津大剧院的大屋顶覆盖的广场，也可以是办公建筑中的中庭、回廊。这些附加的空间并不是最初的设计要求，却可以显著提高建筑的品质。

周： 请问马院长，gmp对于现代性的理解与中国本土建筑师的理解有何不同？

马： 中国本土建筑师受环境影响更多，会较易屈服于业主、政府的意志和干预。除非到了很高的层次，大部分建筑师很难坚持自身的独立追求，个性化的建筑风格很难实现。中国本土建筑师，一方面背负了中国历史传统的传承责任，另一方面又必须面对时代创新追求的压力，对此，他们尚在探索阶段，还未找到平衡点。相对而言，以gmp为代表的德国建筑师对于现代主义建筑精神的把握更为准确。gmp很严肃地对待环境，对现代主义的理解也更纯粹，他们没有中国建筑师那么强烈的负重感和压力，但对于处理历史传承和创新之间的关系，他们更加游刃有余，也更实际一些。gmp在建筑空间、形式、材料上的追求一方面来自于他们在建筑设计方面积累的功力，但建筑师团队自身的理想坚持也功不可没。

周： 请胥茨先生谈一下您对于德国当代建筑的评价，中德建筑师可以有哪些方面的交流与合作？

胥茨： 前面已经说过，德国对新建筑的需求非常少。因此德国建筑师所面临的处境与中国建筑师完全不同。德国建筑师所要面对的挑战是如何在现有的城市语境中利用现有资源对现有建筑进行改造，如何建设智能环保型建筑，使其长久地可持续运营。在这一方面，中德两国有良好的未来合作机会。如何尽量节能环保地进行城市建设，如何合理规划基础设施，使城市、交通与建筑拥有长久的生命力，是中国今后建筑发展所面临的挑战。而关于现代与传统如何协同作用的问题，我相信中国人将会逐渐减少对建筑象征性与隐喻性的关注，而更多着眼于对现有建筑与城市的优化利用。

周： 马院长，您和gmp已经合作了近20个项目，合作过程中贵院主要有哪些收获？

马： 主要有两个收获，一是gmp倡导的对话建筑学，这点使我本人和我们整个建筑师团队受益匪浅。在一个项目进行的过程中，首先要将设计理念和高品质的实现作为始终贯彻的目标。一方面需要听取多方面意见，另一方面也要有能力坚持自己的想法，最终达成共识。这是一种关于工作方式的收获。国内建筑师容易走两个极端，要么固执己见，要么轻易妥协，这对自己和业主都是不负责任的。有效沟通的目的就是促成建筑师自我和业主双方目标同时最大化地实现。

第二个收获是关于全专业化。我院发展经历了三个发展历程，第一个是80年代与德国建筑

It is our objective to create spaces that are used by people and where people feel at home. This is why we have extended the original briefing in almost all projects to the effect that we also create as large and diverse a range of public and semipublic spaces as possible, where people can meet and find agreeable conditions for human interaction. These spaces can have different dimensions and meaning, such as the Grand Forum in the National Museum, the spaces beneath the roofs of the opera houses in Qingdao or Tianjin, or the large forums and galleries in our office buildings. These spaces were not an original part of the briefing. They are additions, but they are additions which—to a large extent—determine the quality of the project.

Zhou Rong   Mr. Ma, how does the understanding of "modern architecture" differ between German and Chinese architects?

Ma Lidong   I feel that Chinese architects are more influenced by their political environment or the building owners. This is an impediment to many Chinese architects in reaching a higher level. But if they do not reach that, then it becomes difficult to develop their own design idea and present an independent style. Of course, Chinese architects have the responsibility to extend Chinese tradition and to expand the cultural understanding associated with it. At the same time, in terms of their creativity, they need to keep abreast of modern developments. At this point in time, they are still trying to find their way; there is not yet a proper balance between the two extremes. I have no doubt that German architects, such as those from gmp, can better understand and interpret the term "modern architecture" than their Chinese colleagues. For a practice such as gmp, it is much easier to reflect upon the relationship between history and modern creativity, and to put the result of this reflection into practice. Thanks to its experience and competence, gmp can apply its own principles with respect to space, form, and material. Therefore, gmp's insistence on their design idea was the key to success.

Zhou Rong   Mr. Schütz, how do you assess contemporary German architecture and the possibility of an exchange between German and Chinese architecture?

Stephan Schütz   As I have said before, the situation faced by German architects is quite different to that of our colleagues in China. In Germany, we are faced with the task of finding out how existing resources in the city can be used and converted, and how sustainable buildings should be built. I believe that in the field of architecture, there is potential for a German/Chinese exchange regarding this subject in the years to come. How is it possible to create infrastructure buildings with appropriate long-term function on the basis of what is existing? The question as to how tradition and the Moderne can interact in China will focus less and less on symbolism and metaphors in buildings, and instead on making use of what already exists.

Zhou Rong   Mr. Ma, you have completed over twenty projects in cooperation with gmp. What experience have you gained?

Ma Lidong   There are mainly two aspects I would like to mention. The first has to do with "dialogue in design." We all have learned a lot from that—our practice team, and I myself too. For example, that—on the one hand —one must keep one's faith in the design concept during the design and implementation phase, and one must insist on its high-quality implementation. On the other hand, one has to remain open to proposals from all sides. But until a consensus of all opinions has been found, one should insist on the design concept as a matter of principle. To learn how good it is when an architectural practice works in this way - that was a very positive experience. Because Chinese architects all too often adopt one of two extreme positions: they are either too stubborn and headstrong, or they do the opposite and make far too many poor compromises. Such behavior is not responsible, because it is detrimental to both the building owner and the architects themselves. By contrast, good and effective communication offers the opportunity to implement the architect's design idea as planned, and to integrate the building owner's ideas during this process.

师合作的燕莎的项目，第二个是90年代与贝聿铭事务所合作的中国银行项目，第三个就是与gmp合作的国博项目。

这三个项目有相同之处，都是在建筑师统领之下全专业的高度协同设计，实现建筑作品功能、技术和艺术的高度统一。

而国博不仅将我院自身的所有专业相结合，把其他研究所（国家队）也调动起来，例如地基、大跨结构、防火等方面，全面协调，最终实现项目功能、技术、艺术、经济性多方面的统一，这使得我们院全专业的协作意识被大大调动起来了。

周： 能否请胥茨先生展望一下gmp未来在中国的发展前景？

胥茨： 随着中国经济和社会文化的持续发展，一种理性的思维模式将会逐渐成为主流。大型的形象工程、地标性建筑将日渐式微，建筑的新出发点将更加关注普通人的生活，建筑规模也会充分考虑周围环境的需求。建筑形式也将更趋于理性化，更加关注环境和公共场所的建设，使普通民众能够享受到更多更好的公共空间，这是gmp在未来中国项目建设中所关注的主要方面。我们坚信，gmp在中国将拥有更加成功的未来。我们诚挚希望能与中国同仁一起创造更多节能环保、舒适宜人的新建筑。

The other important experience concerns the professionalism of the design processes. When you look back over the history of our practice cooperation, you can identify three distinct phases. The first phase is that of the cooperation with German architects on the Lufthansa Center in Beijing project in the 1980s. The second is the cooperation with I.M. Pei on the Bank of China Tower in the 1990s. And the third phase was characterized by the cooperation with gmp on the extension of the Chinese National Museum.

All three projects have in common the holistic design approach under the leadership of the architects. In all cases, this was the prerequisite for being able to satisfy the demanding brief with respect to function, technology and aesthetics.

But the complexity of the design process for the museum was yet another level of experience. I've mentioned it already: this project required not only the involvement of all departments in our practice, but also the specialist teams of the State Institute, foundation experts, structural experts for the construction of large spans, fire safety experts, etc. The cooperation of all these experts was needed to ensure that the project could be completed to this high degree of quality in terms of function, technology, aesthetics and—not least —cost efficiency. And that provided extra motivation to all our members of staff to give their very best.

Zhou Rong   Mr. Schütz, how do you assess the gmp's future in China?

Stephan Schütz   Owing to the economic conditions—which are also changing in China—more rational approaches will gain ground, which are less focused on the demonstration of importance and commercial power, and more on the actual requirements of the people. I believe that the time of large, symbolic projects will come to an end. This move towards a rational architecture, which tackles the enormous environmental problems, as well as the ubiquitous desire to create public spaces where people feel at their ease, is a great challenge in which gmp will make its contribution.

We believe that gmp will have a successful future in China. We hope that—in cooperation with Chinese colleagues—we will be able to complete many interesting architectural projects that will contribute to shaping the architecture of this country in a socially responsible way, which is both sustainable and aesthetically pleasing.

中国国家博物馆坐落于天安门广场西侧，面对人民大会堂，
背景为故宫博物院

中国国家博物馆坐落于天安门广场西侧，面对人民大会堂，
背景为故宫博物院
Photomontage of the National Museum of China situated at the
Tiananmen Square, opposite the Great Hall of the People, in the
background of the ForbiddenCity.

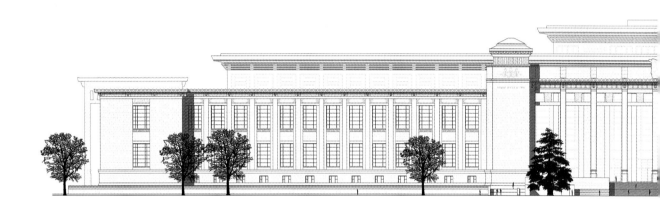

西立面
West elevation

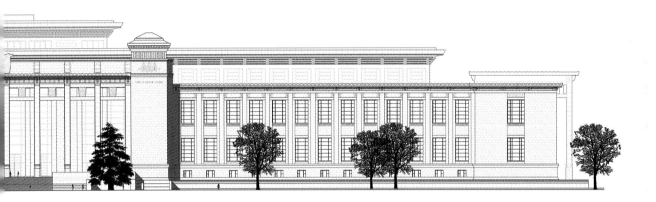

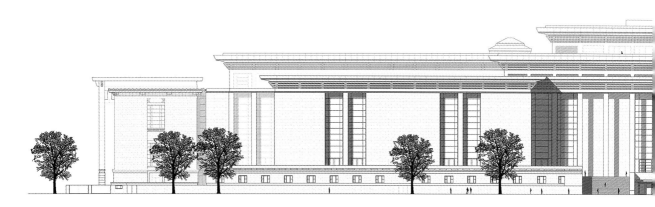

东立面
East elevation

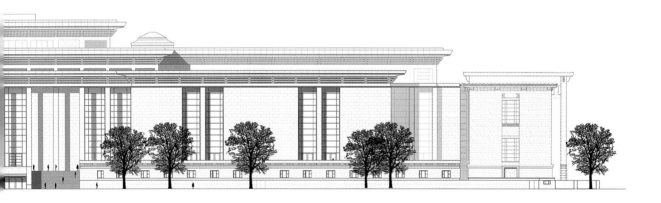

0   5   10   20   50m

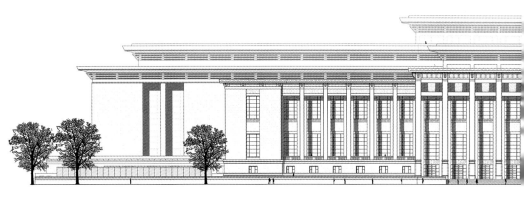

0　5　10　　20　　　　　　50m

北立面
North elevation

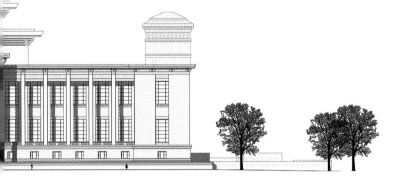

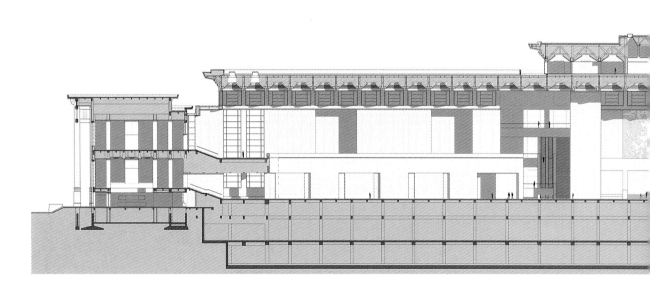

西大厅纵向剖面
Forum: Longitudinal section

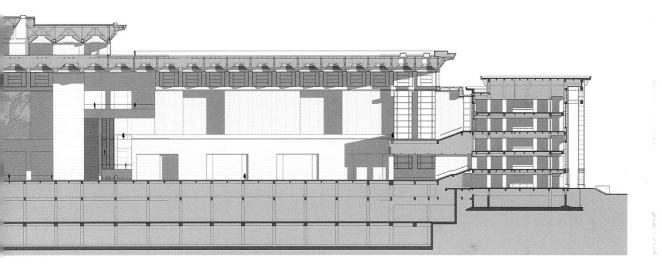

0   5   10   20   50m

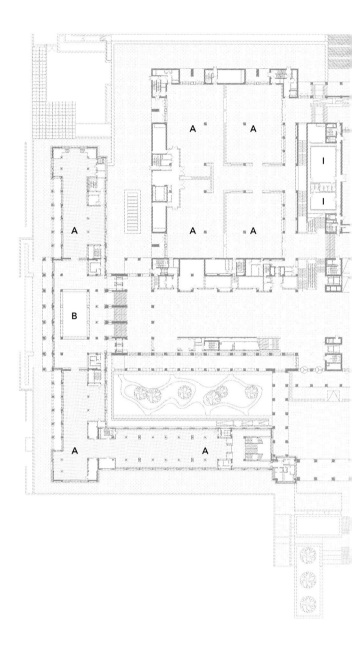

↑ 重点平面层、主入口、西侧入口层面
Floor plan of Main Level - Forum,
Main Entrance, Entrance Level west
A 陈列厅 Exhibition
B 北侧前厅 Northern Foyer
C 图书馆 Library

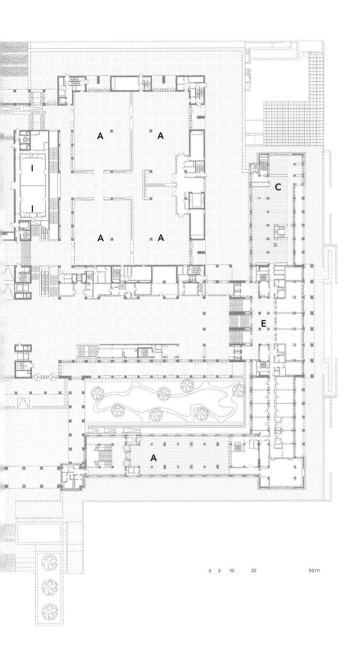

0 5 10   20   50m

D 博物广场 Grand Forum
E 南侧前厅 Southern Foyer
F 西入口前厅 Courtyard West Entrance
G 中央大厅 Central Hall
H 东侧前厅 Eastern Foyer
I 接待室 VIP Reception Rooms

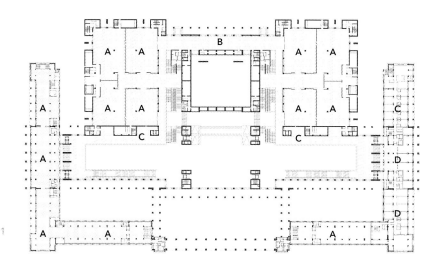

1

0   50m

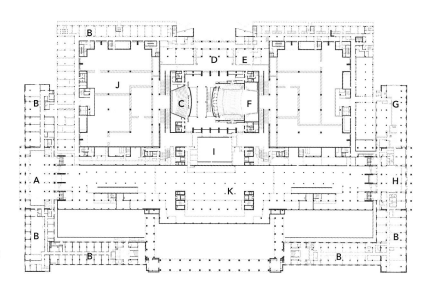

2

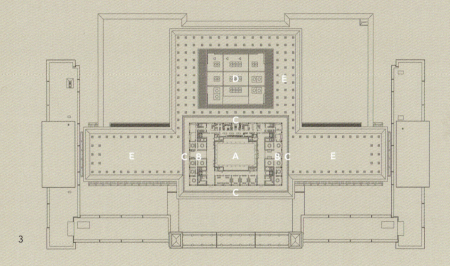

↑ 3
屋面层平面
Floor plan Roof level
A 玉厅 Jade Hall
B 贵宾餐厅 VIP Restaurant
C 屋面平台 Roof Terrace

D 下沉式设备间
   Technical Engineering (lowered)
E 绿化屋面 Green Roof

摄影图集　　　　　　　　　　　　Photographic Documentation

↗ 东立面细部
Detail of the eastern façade

→ 天安门广场一侧入口
右: 新建筑屋顶退阶设计
Entrance area at the Tian'anmen Square.
Right side - the building with the inset
staggered roofs of the new building.

↗ 老国博建筑极富特点的柱廊和开窗样式通过当
代的建筑语言得到了重新演绎
The colonnades and window styles - typical
of the existing building - were translated into
contemporary design vocabulary.
← 东立面细部
Detail of the eastern façade

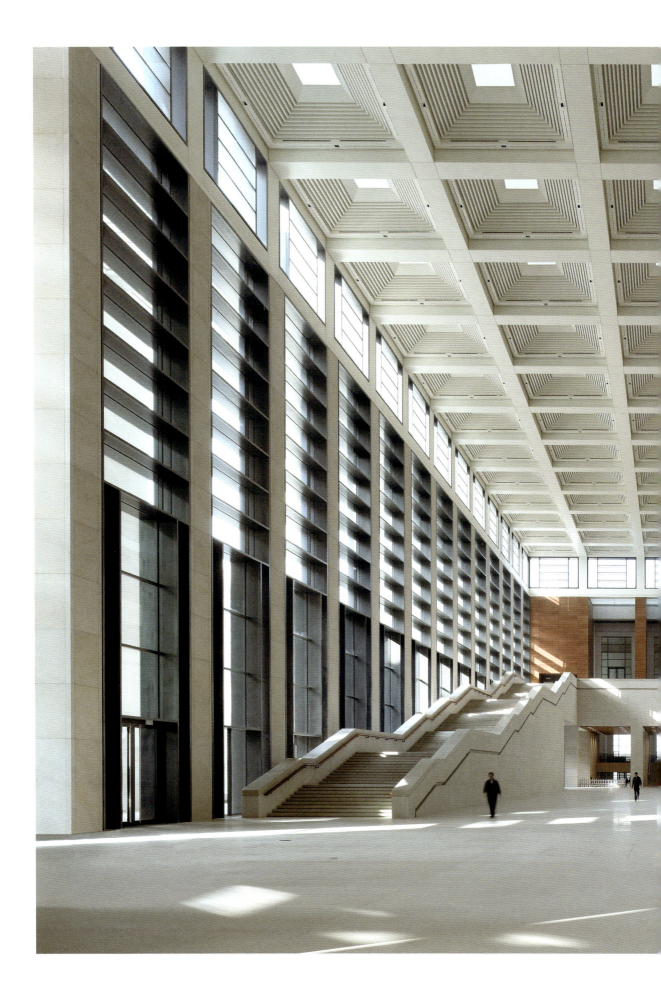

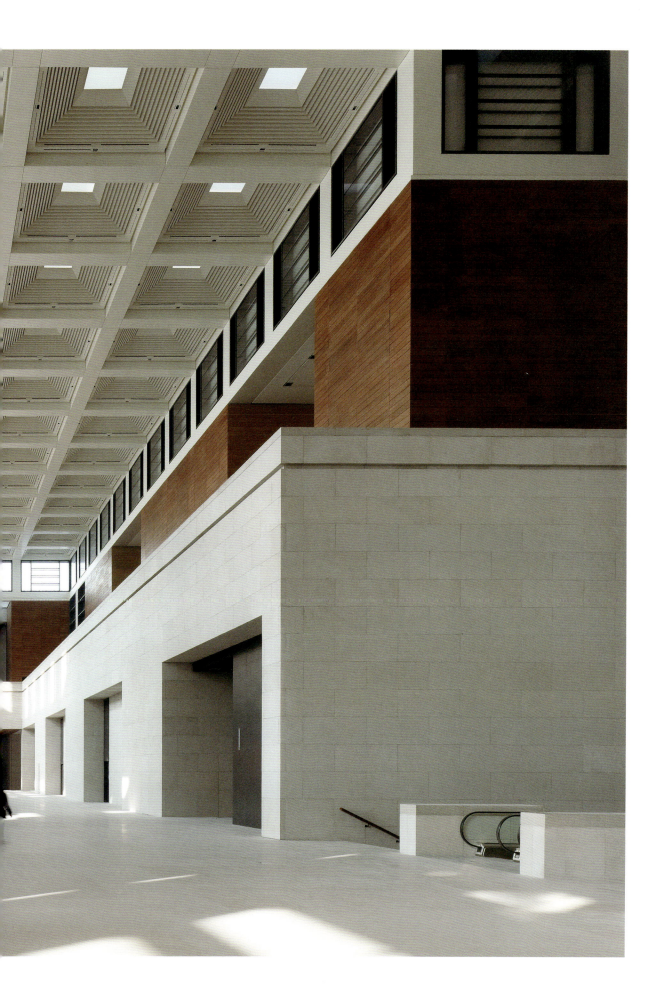

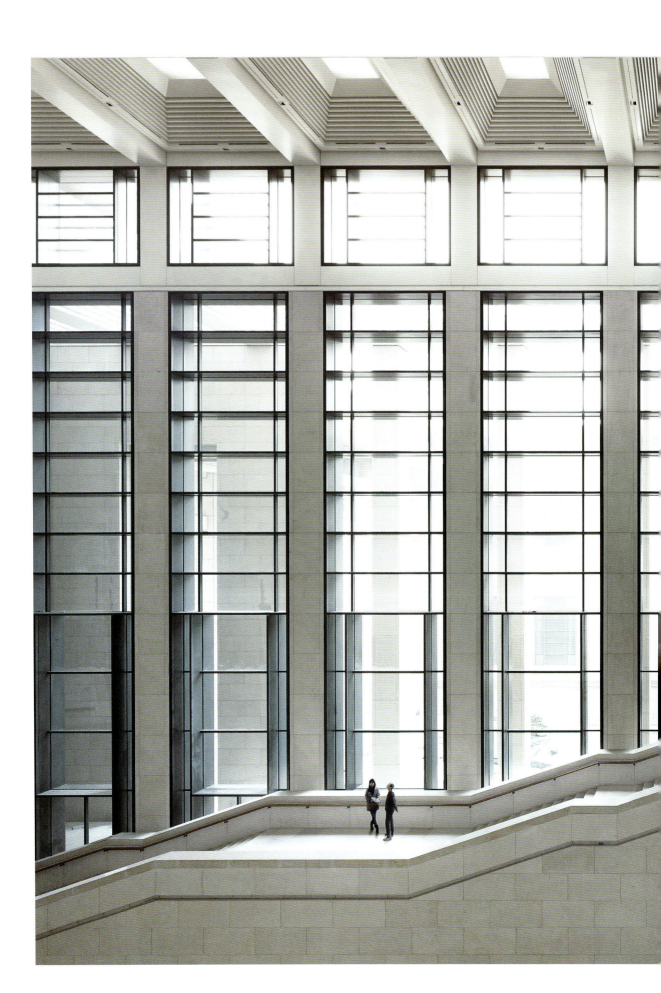

□ "博物广场"面向
西侧视野
Western view inside the
Forum

←← □ "博物广场"面积约9 000m²
The Grand Forum is close to
9,000m²

← □ 西大厅内的巨大阶梯
Staircase inside the
Forum

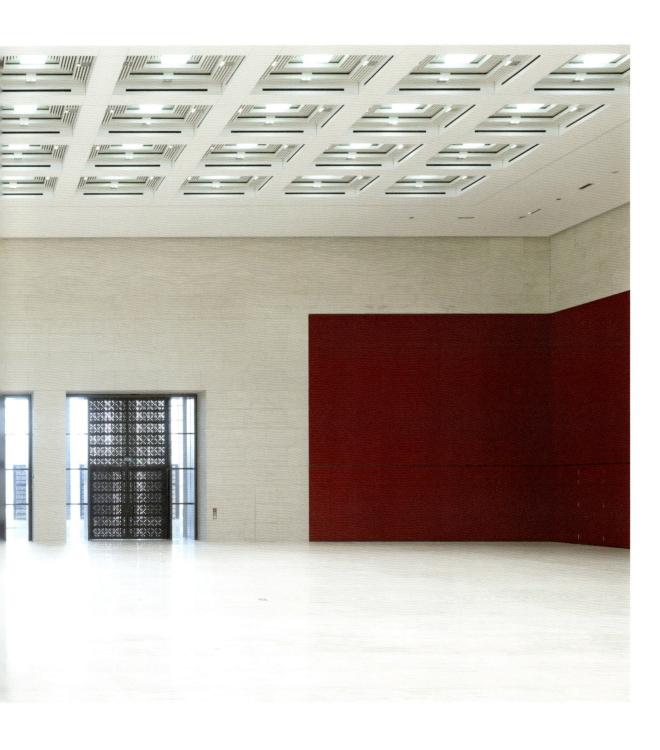

↑   中央大厅以红色墙幔进行装饰
The walls of the Central Hall
are covered with red fabric.

↑→   "博物广场"参观者界定方向，并连接馆内
       的所有公共空间
       The Forum helps visitors with orientation;
       all public areas of the building are acces-
       sible from here.

↑→   "博物广场"参观者界定方向，并连接馆内
       的所有公共空间
       The Forum helps visitors with orientation;
       all public areas of the building are acces-
       sible from here.

↑　西大厅之上的"玉厅"，背投光源经过可回收玻
　　璃饰面板过滤，营造出独特的室内氛围
　　The Jade Hall above the Forum is notable
　　for its backlit cast glass plates made of
　　recycled material.

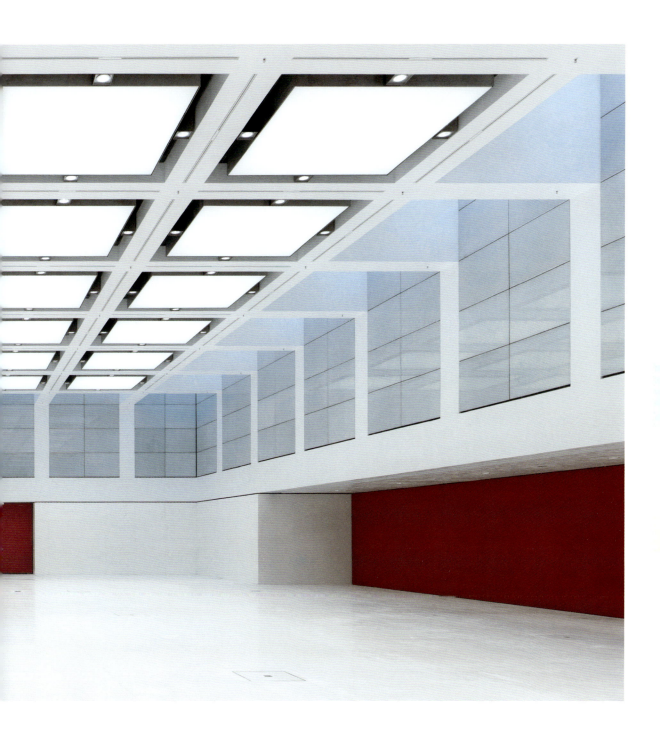

↑   建筑内部采用了产自中国的浅色花岗岩石材
      搭配樱桃木墙面
      Aside from light Chinese granite stone,
      black cherry cladding was used inside the
      building.

↗   入口大门和馆内护栏上采用了
      中国传统的镂空纹饰
      The ornamentation of the entrance doors
      recurs in the forming of the balustrades in the
      interior of the museum.

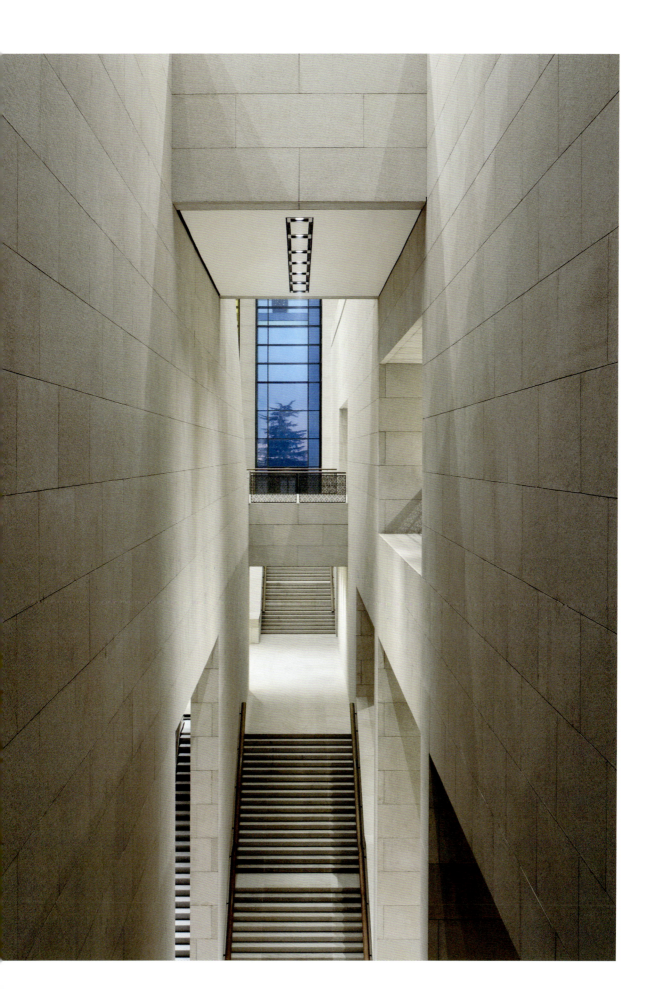

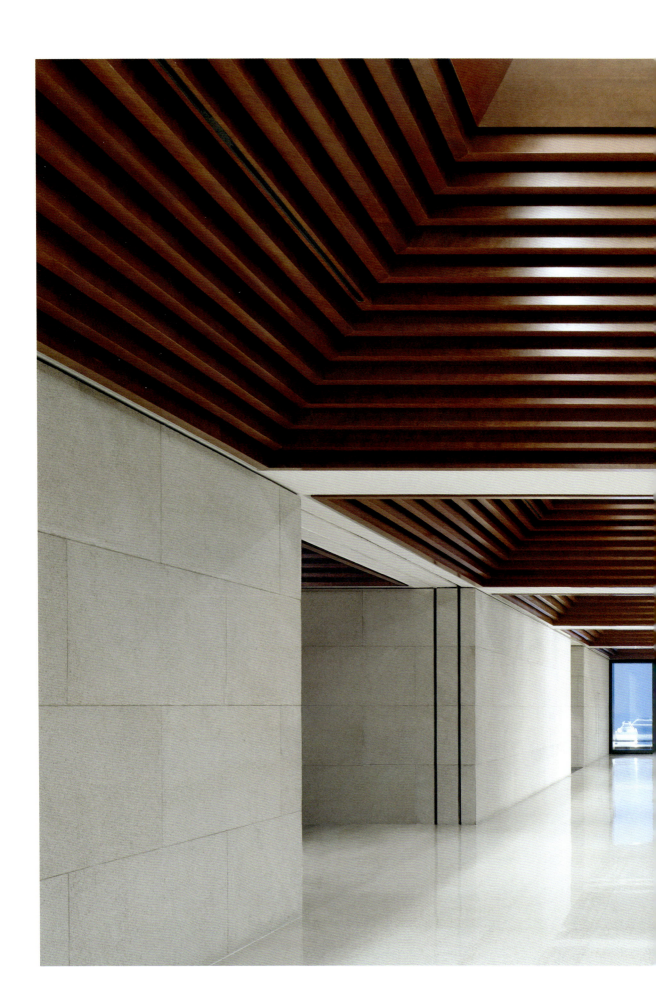

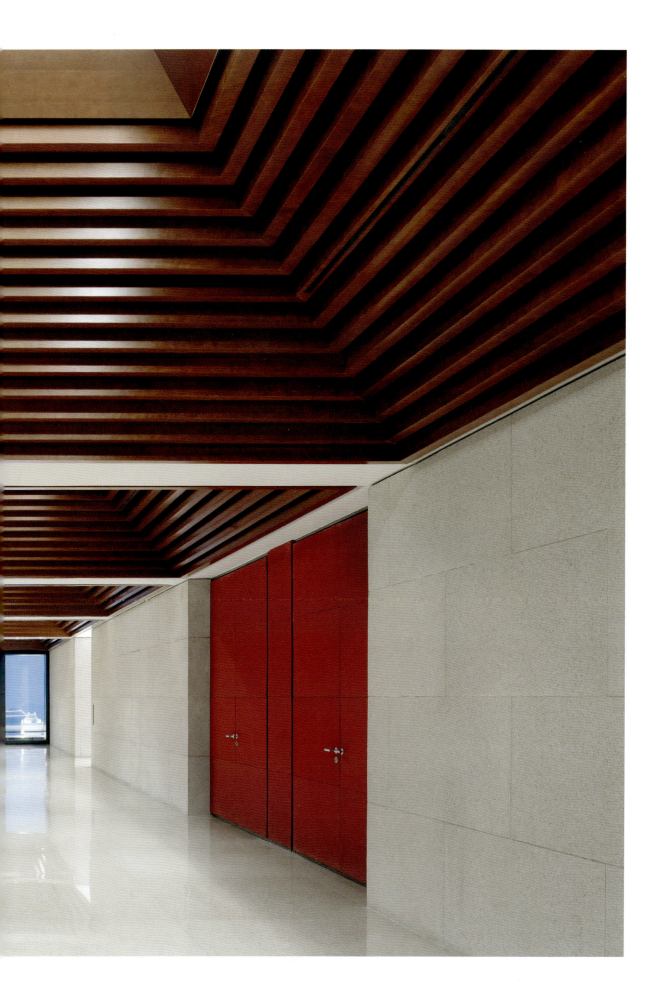

← ▢▢ 通往屋面平台的走廊
Corridor with access
to the Roof Terrace and
Jade Hall

← 通往贵宾接待室的走廊
Corridor to the VIP
Reception Rooms

↑ 临近屋面平台的贵宾区餐厅
Main Dining room in the
VIP area close to the Roof
Terrace

↑   中央大厅之下的学术报告厅，设有264个座席
The auditorium below the Central Hall
offers 264 seats.

↑   国博剧院拥有714个座席
The theater has 714 seats.

东侧入口处的前厅为会议厅以及学术报告厅服务
The lobby area for the Auditoriumand
Theater at the eastern entrance

↑   图书馆的阅读区同样采用了樱桃木墙面板
Also in the reading area of the library, the walls
were clad with black cherry wood.

→   老国博图书馆
Library in the old building

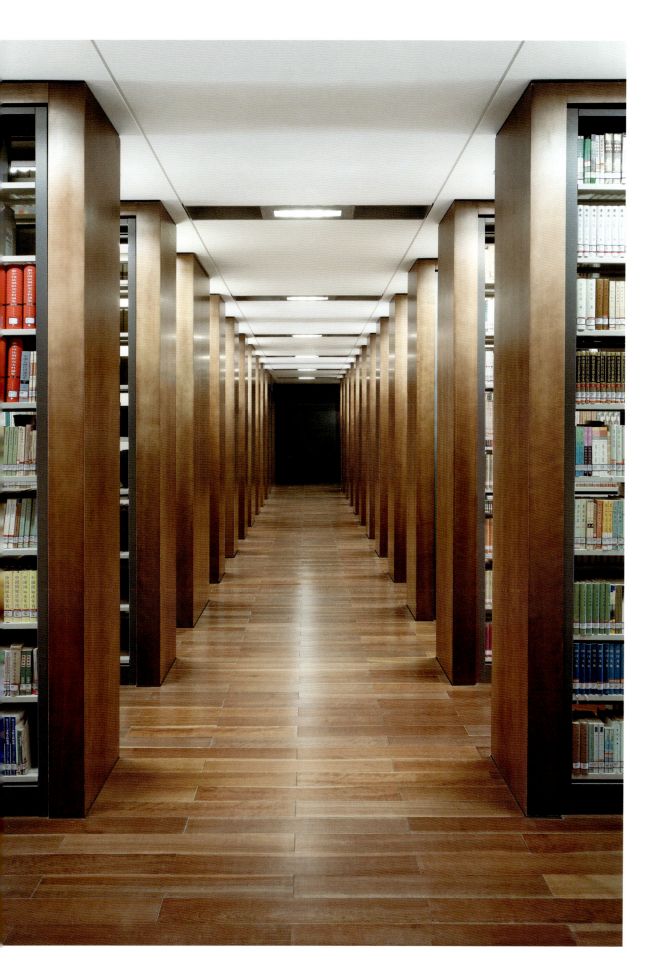

↖↑   地下层 "古代中国" 陈列厅
'Ancient China' exhibition space in the basement

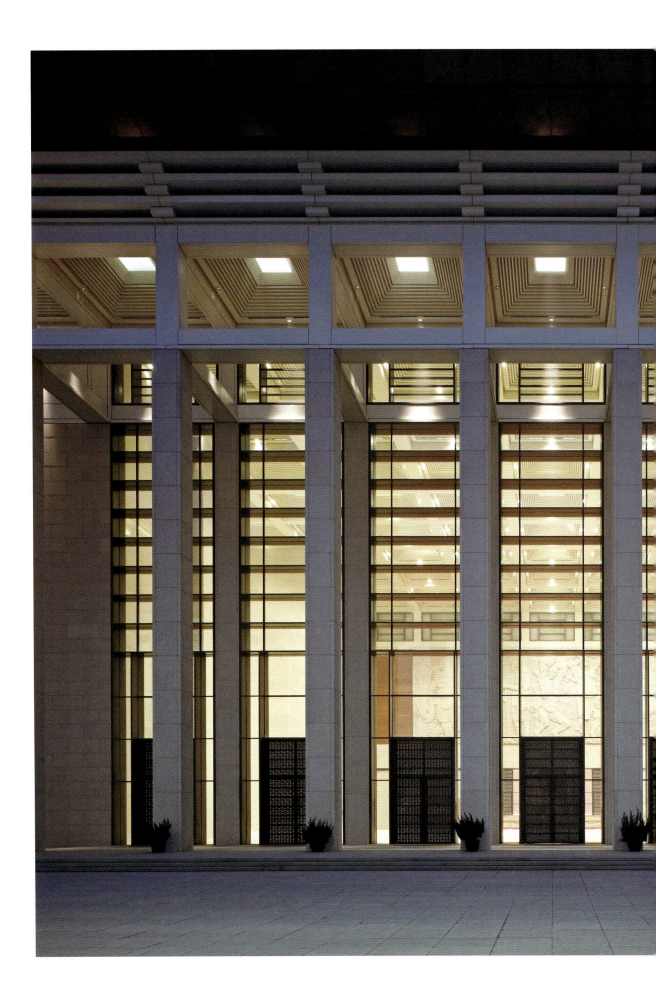

西侧入口
Western entrance

附录

Appendix

德方设计团队 Design Team

曼哈德·冯·格康
教授，荣誉博士，工程硕士，德国建筑师协会建筑师（BDA）

创始合伙人，1935年出生于拉脱维亚里加，1965年作为创始合伙人与福尔克温·玛格共同创立冯·格康，玛格及合伙人建筑师事务所；1974年受聘于布伦瑞克大学任正教授，主持A设计教研室，任建筑设计A研究所所长；受聘于汉堡高等自由艺术学院；2005年，获台湾中原大学设计系荣誉博士学位；2007年，任中国上海华东师范大学设计学院荣誉教授；2007年，任建筑文化学院（aac）院长；获奖众多，其中包括德国弗里茨·舒马赫建筑奖、罗马尼亚国家奖、汉堡高等自由艺术学院奖章；德国建筑师协会（BDA）大奖，并被授予德意志联邦共和国一等十字勋章。

Meinhard von Gerkan
Prof. Dr. h. c. mult. Dipl.-Ing. Architekt BDA

Founding partner, born 1935 in Riga/Latvia; 1965 co-founder with Volkwin Marg of the architectural partnership of von Gerkan, Marg and Partners; 1974 appointed professor, chair A for Design at the Carolo-Wilhelmina Technical University in Brunswick, head of Institute A for Architectural Design, member of the Freie Akademie der Künste in Hamburg; 2005 honorary doctorate in design from Chung Yuan Christian University in Chung Li/Taiwan; 2007 honorary professor, East China Normal University College of Design, Shanghai/China; 2007 president of the Academy for Architectural Culture (aac). Numerous awards, including Fritz Schumacher Prize, Romanian State Prize, bronze plaque of the Freie Akademie der Künste, Hamburg, BDA Prize, Federal Cross of Merit.

斯特凡·胥茨
工程硕士，德国建筑师协会建筑师（BDA）

合伙人，1966年出生于德国杜伊斯堡，2006年成为冯·格康，玛格及合伙人建筑师事务所合伙人，gmp柏林、北京、深圳分部负责人，代表作品有：魏玛新会议中心、柏林新娱乐中心、北京海淀基督教堂、北京中青旅大厦、青岛大剧院、中国国家博物馆、深圳体育中心。

Stephan Schütz
Dipl.-Ing. Architekt BDA

Partner, born 1966 in Duisburg, graduate architect; partner at von Gerkan, Marg and Partners since 2006; management of the gmp offices in Berlin, Beijing and Shenzhen; projects include New Weimar Hall, New Tempodrom, Berlin, Christian Church, Beijing, CYTS Plaza, Beijing, Qingdao Grand Theater, National Museum of China, Beijing, Shenzhen Universiade Sports Center.

gmp

多莉斯·谢夫勒

工程硕士，德国建筑师协会建筑师（BDA）

柏林自由执业建筑师，1967年生于慕尼黑，1994年以优异成绩毕
业于布伦瑞克工业大学，1994年起就职德国冯·格康，玛格及合
伙人建筑师事务所柏林分部，2000—2004年任gmp北京分部
负责人，2009年于柏林建立独立事务所。代表作品为：VAU餐厅
设计、文艺复兴火车站、魏玛会议中心、KPM皇家瓷器作坊改
建、天津生物技术园、Diensdorf度假屋（20世纪获德国木结构
建筑奖）、柏林格吕瓦尔德住宅、柏林Ruhleben区30年代别墅
的改建。

Doris Schäffler

Dipl.-Ing. Architekt BDA

Freelance architect in Berlin, born 1967 in Munich;
diploma with distinction from Braunschweig Technical University 1994; working for von Gerkan, Marg
and Partners Berlin since 1994, from 2000 to 2004 in
charge of gmp Peking, own practice in Berlin since
2009. Projects include Restaurant VAU, "Renais-
sance of railway stations," Weimarhalle, KPM, CYTS,
Tianjin Biotech, Diensdorf Summer House (received
German Timber Construction award), residence
Berlin-Grunewald, conversion of 1930s style villa
Berlin-Ruhleben.

施蒂芬·瑞沃勒

建筑师，工程硕士

项目合伙人，1967年生于丹麦斯基沃，2010年成为德国冯·格康，玛
格及合伙人建筑师事务所项目合伙人，gmp北京分部负责人，担任
北京中国国家博物馆、天津西站和广州电视中心等项目负责人。

Stephan Rewolle

Dipl.-Ing. Architekt

Associate partner, born Skive/Denmark 1967;
graduate architect, associate partner at von Gerkan,
Marg and Partners since 2010; management of
gmp's Beijing office; project management of
Chinese National Museum/Beijing, Tianjin West
railway station, TV Station/Guangzhou etc.

中国区合伙人 gmp Partner China

吴蔚
建筑师, 瑞士联邦高等工业大学(ETH)工程硕士

中国区合伙人，1971年生于甘肃兰州，1992年毕业于重庆建筑工程学院建筑系。1992—1994年于甘肃省建筑勘查设计院工作，1994—1995年赴瑞士学习德语，1997年于瑞士联邦高等工业大学（ETH）建筑系学习。1997—2000年在瑞士苏黎世市Skyline Architecture建筑师事务所及Swiss German咨询顾问公司任职。2000年师从马里奥·坎比教授获得获得瑞士联邦高等工业大学建筑学硕士学位（Diplom）。2000年于德国冯·格康，玛格及合伙人建筑师事务所任职，2001年起任gmp北京、上海分部首席代表，2004年起任gmp事务所项目合伙人，2009年成为gmp中国区合伙人。

Wu Wei
Dipl. Architekt ETH

Partner for China, born 1971 in Lanzhou/Gansu, China; completed architectural studies in Chongqing, China, in 1992. From 1992 to 1994 architect at the Gansu Institute of Architectural Design, 1994 to 1995 German language studies in Switzerland, 1995 to 1997 studied architecture at the ETH Zurich and from 1997 to 2000 worked for Skyline Architecture and SwissGerman Consulting in Zurich; qualified at ETH Zurich (Prof. Mario Campi) in 2000, since then work for von Gerkan, Marg and Partners, Architects (gmp); since 2001 chief representative of gmp's offices in Beijing and Shanghai, since 2004 Associate Partner with von Gerkan, Marg and Partners and since 2009 Partner at gmp with responsibility for China.

中方设计团队 Partner Practice   CABR

马立东

中国建筑科学研究院副总建筑师，建筑设计院院长，国家一级注册建筑师

1969年生于北京，1995年毕业于清华大学建筑学院，先后获得该校建筑学学士和工学硕士学位。曾先后主持多项国家重点建设工程及其他具有重大影响的建筑工程设计项目，包括北京盛福大厦、科威特大使馆及官邸、北京SOHO尚都、中国疾病预防控制中心、中国国家博物馆改扩建工程的设计工作。

Ma Lidong

Director of Architectural Design Institute
Deputy Chief Architect
China Academy of Building Research
1st Class Registered Architect, PRC

Born in 1969 in Beijing, graduated from the School of Architecture of Tsinghua University, Master degree. Deputy Chief Architect of China Academy of Building Research (CABR), the Director of the Architectural Design Institute of CABR; Projects include the Beijing Sunflower Tower, the Kuwait Embassy in China, the SOHO Shang-Du Complex Beijing, Chinese Center for Disease Control and Prevention, and the National Museum of China.

王双

中国建筑科学研究院建筑设计院副总建筑师，国家一级注册建筑师

1968年生于北京，1992年毕业于清华大学建筑学院，同年进入中国建筑科学研究院，曾作为项目负责人主持过多项国家重点建设工程及其他具有重大影响的建筑工程设计项目，包括：北京京西学校、中国银行总部大厦、印度驻华使馆、华润清河五彩城、中国国家博物馆改扩建工程、珠海博物馆和规划展览馆、达美中心广场等。

Wang Shuang

Deputy Chief Architect
Architectural Design Institute
China Academy of Building Research
1st Class Registered Architect, PRC

Born 1968 in Beijing. 1st Class Registered Architect. Graduated 1992 from School of Architecture, Tsinghua University. Deputy Chief Architect of ADI, CABR; Project includes: International School WEB; Bank of China headquater Bejing; Office Building Renovation Ministry of Construction; Embassy of India, Beijing; Dreamport of C.R.Land, Beijing; National Museum of China; Zhuhai Museum; Damei Plaza.

项目负责人 Project Management

马蒂亚斯·魏格曼

建筑师，工程硕士（应用技术大学）

设计总监，1969年生于德国沃尔姆斯。2010年起就职德国冯·格康，玛格及合伙人建筑师事务所北京分部，负责项目有中国国家博物馆、上海外滩SOHO、上海复兴路SOHO、天津国家会展中心等。

Matthias Wiegelmann

Dipl.-Ing. Architekt (FH)

Born 1969 in Worms/Germany. Director at von Gerkan, Marg and Partners since 2010, member of gmp Peking office; project management includes: Chinese National Museum, Peking; Shanghai SOHO Bund; Shanghai SOHO Fuxinglu; National Exhibition Centre, Tianjin.

帕特里克·弗莱德勒

建筑师，工程硕士

项目设计总监，1967年出生于德国图宾根，2012年成为德国冯·格康，玛格及合伙人建筑师事务所项目设计总监，2005—2012年于gmp北京分部工作，中国国家博物馆施工阶段项目负责人，代表作品有：江苏淮安国际会展中心、广州赫基国际大厦。

Patrick Pfleiderer

Dipl.-Ing. Architekt

Born 1967 in Tübingen/Germany, Associate Director at von Gerkan, Marg and Partners since 2012, member of gmp Peking office from 2005 to 2012, project management includes: Chinese National Museum Peking (execution), Huai An International Exhibition Centre, Jiangsu Province, Trendy International, Guangzhou.

作者和摄影师 Authors and Photographers

吕章申
中国国家博物馆馆长

吕章申，1955年12月生，现任中国国家博物馆馆长、中国博物馆协会名誉理事长、中国建筑文化研究会会长、中国对外文化交流协会副会长、中国书法家协会理事、中国书法院特邀研究员、中国国际书画协会顾问、中国收藏家协会顾问。其书法作品在多种报刊、杂志发表，并被多家博物馆、纪念馆等收藏。曾出版《吕章申书法集》《吕章申书法作品集》。

Lv Zhangshen
Museum Director of the Chinese National Museum, Beijing

Born in December, 1955. Director of the National Museum of China. Vice Chairman of the China International Culture Association,Director of the China Calligraphers Association, special researcher with the Chinese Academy of Calligraphy and consultant to the China International Association of Calligraphy and Painting and the China Association of Collectors.Lv's works of calligraphy have been published in newspapers and magazines, and are part of museum and memorial hall collections. Other publications include: Collection of "Lv Zhangshen's Calligraphy (2005)" and Collection of "Lv Zhangshen's Calligraphy Works (July 2007)".

马丁·罗特
博士、教授

1955年生于斯图加特，在图宾根埃伯哈特－卡尔斯大学攻读博士，博士论文主题为文化史博物馆的历史。1987—1988年，在巴黎的人文科学之家和德国历史研究所研究博物馆与世界博览会的发展史及其展览案例。1991—2000年，担任德累斯顿德国卫生博物馆馆长。1996—2001年，参与2000年汉诺威世博会的管理，负责主题公园、全球性项目和全球对话活动等事务。1995—2003年，担任柏林德国博物馆联合会主席。2003年起，担任德累斯顿理工大学文化政治学和文化管理学荣誉教授。2001—2011年，担任德累斯顿州立美术馆馆长。2011年起，担任伦敦维多利亚和阿尔伯特博物馆馆长。

Prof. Dr. Martin Roth

Born 1955 in Stuttgart, awarded doctorate by Eberhard-Karls University in Tübingen for his dissertation on the history of museums of cultural history. 1987 to 88, research at the Maison des Sciences de l'Homme and the Institute of German History in Paris on the subject of the history and practice of world exhibitions and museums, amongst others. 1991 to 2000, curator of the German Hygiene Museum, Dresden. 1996 to 2001, member of EXPO 2000 management, with responsibility for the theme park, the worldwide projects, and the global dialogue at EXPO 2000 GmbH, Hanover. 1995 to 2003, president of the German Museums Association, Berlin. Since 2003, honorary professor for cultural policy and culture management at Dresden Technical University.
2001 to 2011, director General of State Art Collections in Dresden and since 2011, Director of the Victoria&Albert Museum in London.

安德拉斯·桑托
博士、作家和记者

《艺术新闻报》特约编辑，同时也是"艺术世界沙龙"的联合创始人。《纽约时报》《艺术论坛》《博物馆实践》和很多其他出版物都刊登过他的文章。他还在苏富比艺术学院教授艺术商业和艺术营销的课程，并且是哥伦比亚大学新闻学院主任。作为一名顾问，他为世界各地的博物馆、基金会提供咨询，同时还主持了巴塞尔艺术对话系列节目。

Dr. András Szántó
Author and Journalist

Contributing editor for "The Art Newspaper" and co-founder of "ArtworldSalon". His writings have appeared in "The New York Times", "Artforum", "Museum Practice", and other publications. He has lectured in art business and art marketing at the Sotheby's Institute of Art, N.Y., and has been the director of the National Arts Journalism Program and the NEA Arts Journalism Institute at Columbia University. As a consultant, he advises leading museums, foundations, and corporations world-wide. He is a regular moderator of the Art Basel Conversations series.

克里斯蒂安·盖尔
摄影师

1966年生于慕尼黑。1993年至1996年学习建筑学。1996年起作为自由建筑摄影师，为Hild + K、伊东丰雄、墨菲/扬、雷姆·库哈斯、HG Merz等世界著名建筑师事务所拍摄。另外还曾受Keith Sonnier、 Ulrich Rückriem、 Walter de Maria等艺术家以及纽约古根海姆美术馆、纽约现代艺术博物馆、新国家艺术画廊等艺术机构委托进行拍摄。其摄影作品发表于英国《建筑评论》《A+U》《domus》《新苏黎世报》等媒体上。

Christian Gahl
Photographer

Born in Munich in 1966. Studied architecture from 1993 to 1996. Freelance architectural photographer with international commissions for architects Hild + K., Toyo Ito, Murphy/Jahn, Rem Koolhaas, and HG Merz since 1996. His clients also include artists such as Keith Sonnier, Ulrich Rückriem, Walter de Maria, and institutions such as the Guggenheim New York, the MoMA New York, and the new National Gallery in Berlin. His photographs are published in such publications as "The Architectural Review", "A+U", "domus" and "Neue Zürcher Zeitung".

周榕
教授

1968年生，美国哈佛大学设计学硕士，清华大学建筑学院博士，副教授，《世界建筑》杂志副主编，著名建筑评论家。

Prof. Zhou Rong

Born in 1968, graduated from Havard University with an MA in design, Doctor of architecture from Tsinghua University. Vice Professor, Vice chief director of the magazine "World Architecture", renowned architectural critic.

项目信息 Project Data

业主 Client
中国国家博物馆
The National Museum of China

竞赛方案设计 Competition Design
曼哈德·冯·格康、斯特凡·胥茨和施蒂芬·瑞沃勒以及
多莉丝·舍弗勒
Meinhard von Gerkan and Stephan Schütz
with Stephan Rewolle und Doris Schäffler

落实方案设计 Executed Design
曼哈德·冯·格康和斯特凡·胥茨以及施蒂芬·瑞沃勒
Meinhard von Gerkan and Stephan Schütz
with Stephan Rewolle

设计团队 Design Staff
雷格·霍海塞尔、卡特琳·坎努斯、拉尔夫·齐伯、杜鹏、董春嵩
Gregor Hoheisel, Katrin Kanus, Ralf Sieber, Du Peng,
Dong Chunsong

落实方案项目负责人 Proj. leaders Execution
马提亚斯·魏格曼和帕特里克·弗莱德
Matthias Wiegelmann with Patrick Pfleiderer

设计人员 Staff
鲍威、约汉娜·恩辛格、安娜·布兰达、孔晶、
安德拉斯·格茨、郭福慧、穆尔杨托、陈悦、
郑昕、高华、邢九洲、海嘉·雷蒙德、
托比亚斯·凯、克里斯蒂安·多尔恩多夫、
安奈特·吕本、维瑞娜·费什巴赫、姜琳琳、刘焱、
梅拉法林·鲁茨伯里、浦次阳子、韩露、夏琳、
田景海、乌利·巴赫曼、阿加达·古埃尔巴哈、爱瑞斯·柏勒、
萨宾娜·斯德支
Bao Wei, Johanna Enzinger, Anna Bulanda-J., Kong
Jing, Andreas Goetze, Guo Fuhui, Mulyanto, Chen
Yue, Zheng Xin, Gao Hua, Zheng Jiuzhou, Helga
Reimund, Tobias Keyl, Christian Dorndorf, Annette
Loeber, Verena Fischbach, Jiang Linlin, Liu Yan,
Mehrafarin Ruzbehi, Yoko Uraji, Han Lu, Xia Lin, Tian
Jinghai, Uli Bachmann, Ajda Guelbahar, Iris Belle,
Sabine Stage

项目负责人 Project Leaders
马提亚斯·魏格曼和帕特里克·弗莱德
Matthias Wiegelmann with Patrick Pfleiderer

中方合作设计单位 Chinese Partner Practice
中国建筑科学研究院, 北京 (CABR) 马立东, 王双
CABR (Chinese Academy of Building Research), Peking
Ma Lidong, Wang Shuang

结构设计 Structural Engineering
中国建筑科学研究院, 北京 (CABR)
CABR (Chinese Academy of Building Research), Peking

电气设备 Electronic Engeneering
中国建筑科学研究院, 北京 (CABR)
CABR (Chinese Academy of Building Research), Peking

景观设计 Landscaping
雷瓦德景观建筑事务所, 德累斯顿/北京
RLA Rehwaldt Landscape Architects, Dresden/Peking

大型活动工程技术 Event Technology
ADA声学设计事务所, 柏林
ADA Acoustic Design Ahnert, Berlin

照明设计 Lighting Design
观念照明设计有限公司, 德国, 特劳恩施泰因; 清华城市规划设计研究院, 北京
conceptlicht GmbH, Traunreut, Urban Planning Design
& Research Institute of Tsinghua University, Peking

幕墙设计 Facade
SuP工程师事务所, 达姆施塔特/北京
SuP Ingenieure GmbH, Darmstadt/Peking

声学设计 Acoustics
穆勒—BBM集团, ADA声学设计事务所
Müller-BBM Group/ADA Acoustic Design Ahnert

项目管理 Project Management
北京国金工程管理咨询有限公司
Beijing Guojin Consultants Co., Ltd., Peking

建筑面积 GFA 197 113m²

设计开始时间 Initial Planning 2004年

设计调整时间 Replanning 2005年

图片版权 Picture Credits

说明 Imprint

总策划 Concept
迈克尔·库恩
德国冯·格康，玛格及合伙人建筑师事务所（gmp）
公共关系和新闻出版部负责人
Michael Kuhn (gmp)
Head of PR
柳青
《城市·环境·设计》（UED）杂志社执行主编
Liu Qing (UED)
Executive Chief Editor of UEDmagazine

编辑 Editorial Direction
克劳迪娅·苔斯勒 Claudia Tiesler (gmp)
方小诗 Fang Xiaoshi (gmp)
郑珊珊 Zheng Shan Shan (Peking gmp)

平面设计 Layout and Typesetting
汤姆·魏伯伦茨、亨德里克西什莱
欧恩平面设计公司（德国汉堡）
ON Grafik, Tom Wibberenz mit with
Hendrik Sichler, Hamburg

校审 Proofreading
方小诗 Fang Xiaoshi (gmp)
包伸明 Bao Shenming (UED)

翻译 Translation
海德威·布赫 Hartwin Busch, Ashdon, UK (en.)
克里斯蒂安·布莱辛 Christian Brensing, Berlin (de.),
韦劭辰 Wei Shaochen, Grünstadt (chin.)
方小诗 Fang Xiaoshi (chin.) (gmp)
杨凯 Yang Kai, Hamburg (chin.)
李萱 Li Xuan (chin.) (gmp)

图片处理 Picture Editing
特里克茜·汉森 Trixi Hansen (gmp)
吉多·布里克斯纳 Guido Brixner (gmp)

印刷制作 Print Production – Binding
北京雅昌彩色印刷有限公司
ARTRON